LICENSING TECHNOLOGY AND PATENTS

with special reference to the manufacturing and process industries

Vernon Parker

INSTITUTION OF CHEMICAL ENGINEERS

Distributed exclusively in the USA and Canada
by VCH Publishers, Inc, New York

The information in this handbook is given in good faith and belief in its accuracy, but does not imply the acceptance of any legal liability or responsibility whatsoever, by the Institution, or by the author for the consequences of its use or misuse in any particular circumstance.

**Published by
Institution of Chemical Engineers,
Davis Building,
165-171 Railway Terrace,
Rugby, Warwickshire CV21 3HQ, UK.**

**Distributed exclusively in the USA and Canada by
VCH Publishers, Inc,
220 East 23rd Street,
Suite 909,
New York,
NY 10016-4606,
USA.**

Copyright © 1991 Institution of Chemical Engineers

Reprinted 2002

All rights reserved. No part of this publication may be reproduced, stored in a retrieval system, or transmitted, in any form or by any means, electronic, mechanical, photocopying, recording or otherwise, without the prior permission of the copyright owner.

ISBN 0 85295 277 5

Cartoons drawn by Jim Watson.

Printed in England by BPCC Wheatons Ltd, Exeter.

PREFACE

The aim of this book is to tell you what I think licensing technology and patents is all about in a business context. It is a distillation of one man's experience over a period of 25 years in several constituent companies and businesses of the ICI Group, embracing fibres, petrochemicals, plastics, specialities, control systems and latterly explosives and their accessories within ICI Explosives, one of ICI's International Businesses. In that time, I have been as much or more concerned with licensing-in technologies and cooperative developments than with licensing-out, and so you will find the specific and often neglected perspective of the licensee given ample treatment in the text.

The book is aimed first at business men, especially research and technical directors and managers, and project engineering managers. But it is also intended for those amongst my fellow patent attorneys and our lawyer colleagues who are called upon to advise in this area of business activity and who have so far had limited exposure. I trust that its contents will assist each of you to fashion with your associates, clients and advisors more effective deals bringing lasting satisfaction from a standpoint of increased awareness. But it cannot, of course, assure such an outcome. You must weigh up your particular circumstances, be competently briefed, seek and heed the best available professional advice and be skilful in negotiation so that in the end you get what the business needs on acceptable terms and you minimise the scope for unreasonable men later, when faces and corporate positions may have changed, to work the agreements reached against your interests and expectations. With that prudent but necessary disclaimer, read on.

TAKE PROFESSIONAL ADVICE **FIRST!**

CONTENTS

PREFACE   iii

1. INTRODUCTION   1
    THE ESSENTIAL AIM AND LEGAL BASIS FOR LICENCING   1
    INTELLECTUAL PROPERTY DESCRIPTIONS   3
    REGISTERED AND UNREGISTERED TRADE MARKS   10
    SUMMARIES   12
    THE PROTECTION OF CONFIDENTIAL INFORMATION   14

2. THE BUSINESS IMPLICATIONS   18
    THE MOTIVATION TO SEEK OR OFFER LICENCES   19
    THE ADVANTAGES AND DISADVANTAGES TO LICENSORS AND LICENSEES   26

3. TECHNOLOGY LISTING AND EVALUATION   33
    PRE-LICENCE SHORTLISTING OF AVAILABLE TECHNOLOGIES   33
    MANAGING PRE-LICENCE CONFIDENTIAL ASSESSMENTS   34
    CONFIDENTIALITY AGREEMENT TERMS   36
    DESIGN PACKAGE AND LICENCE OPTIONS   40

4. THE COMMON TYPES OF LICENCE AGREEMENT   43
    BARE PATENT LICENCES   44
    SUPPORTED PATENT LICENCES   49
    IMPLIED LICENCES   50
    INFORMATION LICENCES   51
    TECHNOLOGY LICENCES (PATENTS AND INFORMATION)   55
    PAYMENT STRUCTURES OF LICENCES   63
    FOOTNOTE: TECHNOLOGY DEVELOPMENT AGREEMENTS   69

5. THE STRUCTURE AND WORDING OF LICENCE AGREEMENTS   72
    DEFINITIONS CLAUSE   73
    INFORMATION SUPPLY CLAUSES   74
    SERVICES SUPPLY CLAUSES   76

THE GRANTS AND GRANT-BACKS	76
THE PAYMENTS CLAUSES	77
PERFORMANCE GUARANTEES CLAUSE	78
SECRECY CLAUSE	79
VALIDATION, TERM AND TERMINATION CLAUSES	80
THE GENERAL CLAUSES	81
APPENDIX I — INTELLECTUAL PROPERTY RIGHTS	**87**
PATENTS	87
REGISTERED DESIGNS	89
THE UNREGISTERED DESIGN RIGHT	90
EUROPEAN COMPETITION LAW: HOW IT IS STRUCTURED	91
APPENDIX II — THE LICENCE GRANTS	**93**
RESTRICTIONS AND FREEDOMS	93
APPENDIX III — ANNOTATED SAMPLE AGREEMENTS	**102**
SECTION 1: BARE PATENT LICENCE	103
SECTION 2: PATENT LICENCE AGREEMENT	114
SECTION 3: INFORMATION CROSS-LICENCE	124
SECTION 4: JOINT DEVELOPMENT AGREEMENT	133
SECTION 5: PROPRIETARY PRODUCT EVALUATION AGREEMENT	140
SECTION 6: TOTAL TECHNOLOGY LICENCE	145

1. INTRODUCTION

THE ESSENTIAL AIM AND LEGAL BASIS FOR LICENSING

Licensing concerns the supply of useful technical information, by agreement, from some person who has it at his disposal to another person who wants it and can use it commercially. Licensing also concerns, as a distinct and separate aspect, but often linked with the first, authorising others to operate within an area of activity which the State has reserved to a particular person under statutorily created proprietary rights designed to foster progress in the useful arts and applied technology by encouraging creativity and by encouraging the publication of novel discoveries rather than their private working in secret.

An attempted definition of the principal aims of licensing might be:
- to convey from one person ('the licensor') to another person ('the licensee'), in a controlled manner, technical knowledge useful to him within a specified field of manufacture, use and sale of particular products or materials so that he may commercially use that knowledge with the consent of the licensor for such purposes but not otherwise, and/or;
- to authorise a person ('the licensee') to undertake specified activities within a given field of manufacture, use and sale of particular products or materials that, but for such authority, would be an infringement of rights granted by law either to the person giving the authority ('the licensor') or to some other person from whom the licensor derives title. These persons may be individuals or so-called juridical persons, ie firms, companies and corporations.

; The term 'manufacture' is not used in any nar-

'LICENSING CONCERNS THE SUPPLY OF USEFUL TECHNICAL INFORMATION... FROM SOME PERSON WHO HAS IT AT HIS DISPOSAL TO ANOTHER PERSON WHO WANTS IT AND CAN USE IT COMMERCIALLY'

row sense; it encompasses any physical or chemical process or treatment which changes something from one state, form or condition to another so as to add utility or value.

The rights granted by law are, for present purposes: firstly, monopoly rights under patents and, to a lesser extent, under registered designs and registered trade marks, and secondly, rights to prevent unauthorised copying of some physical expression of technology under copyright laws in respect of literary and artistic works and a new UK Design Right. The preferred family name for these statutorily created rights is nowadays 'intellectual property rights'. In the following chapters of this text, only patents are considered in depth because in the generality of cases of technology licensing only patents feature prominently or at all. Why this is so will perhaps be clearer to readers who have not so far had much to do with intellectual property from the explanations of the nature of each type which are included later in this chapter and in Appendix I.

All sorts of technical knowledge are licensable. If it can be passed to another in such a way as to be understood by him, and used by him in manufacture or to enhance his sales, it is licensable.

A general list would certainly include process descriptions, flowsheets, methods, techniques, control systems, computer software, designs, engineering drawings, formulations and specifications and that class of information loosely called know-how which defies precise definition but is a very real element in technology transfer and can considerably shorten the learning process which is involved in any assimilation of new technology. By contrast, personal skills and craftsmanship are not transferable separately from the people possessing them.

It is implicit in technology transfer that the knowledge being conveyed contains what the licensor considers to be confidential, secret and proprietary information. This knowledge will have been gained by the expenditure of time, money and effort and will have called for the exercise of creative and inventive faculties in most instances. It may not be uniquely possessed by the licensor. Others elsewhere may have it, or something similar or which achieves the same ends. Nevertheless, it is a valuable asset in the hands of the licensor and

it will be an important aim of the licensor to arrange the transfer and any ancillary grant of rights under intellectual property in such a way as to ensure there will be a benefit to him and to preserve and maximise, so far as possible, the continuing value of that asset to him. The laws of all jurisdictions will enable him to do this, in particular the laws of patents and designs, the copyright laws, and the laws of contract and confidence. Of course, some legal systems are more developed than others. Civil remedies, and in some cases criminal sanctions, are available which are an effective deterrent to those who might otherwise be inclined to abuse a relationship of trust and confidence, ignore inconvenient undertakings of confidence, misuse trade secrets, indulge in unauthorised use of information in breach of contractual undertakings, or infringe patents, proprietary design rights or copyrights.

Whilst it is convenient to think of confidential information and trade secrets as a species of personal property and to refer, even in agreements, to 'licensing' its use, there are no legal property rights inherent in information or secrets and no monopoly rights as there are with patented inventions and registered designs. The laws of certain States within the USA have come nearest to treating trade secrets as a species of property, with misappropriation of trade secrets being actionable under specific, and increasingly uniform, statutes.

INTELLECTUAL PROPERTY DESCRIPTIONS

PATENTS

A person who makes a new and useful material, substance or chemical compound, or who devises a new and useful machine, piece of equipment, tool or structure, or who discovers a new and useful method of making something, is entitled to seek the grant to him of a patent under national patent systems that exist in most countries around the world. The newness (or 'novelty' as it is usually termed) may be in the nature of a fundamental difference from what has gone before ('the state of the art') or be an incremental, but significant, development from that earlier base. The creative step taken by such person ('the inventor') may be the result of inspiration, insight, 'ringing the changes', or chance.

The first stage in seeking patent protection is critical. A full disclosure of the new invention must be filed at a receiving Patent Office in accordance with the prescribed rules in order to establish a priority date that can be internationally recognised, and this must be done before there has been any

'IF EVERYTHING IS CONFIDENTIAL, NOTHING IS!'

publication or non-confidential disclosure of the invention since, with very few exceptions of local relevance only, novelty of the invention at the priority date is essential for patent validity.

Mere novelty however, is not enough; additionally the subject matter sought to be patented ('the invention') must not be obvious to those who are skilled in the relevant field (and who are presumed to have read up their subject).

The invention must be described adequately and clearly in the patent specification that is eventually to be made available to the public and, equally importantly, the class of things or methods that are to be reserved to the patentee, ie the owner of the patent, for the duration of the patent's life must be delineated in the patent specification by what are called 'Claims'.

This delineation of the sought-for monopoly, in the form of patent claims, is the key determinant of the maximum monopoly rights granted to the patentee. The monopoly ultimately enjoyed by the patentee is frequently much less than what is initially sought or what is first granted by the national Patent Offices after an examination of the invention's merits. The reason for this has to do with the circumstances under which the claims are drawn up — in particular, the patentee's inadequate grasp of all that has gone before ('the prior art'), the fact that since invention lies beyond the frontiers of previous knowledge there can be a lack of real understanding of the factors and forces at play, and there is insufficient time in the rush to get something on file at the Patent Office to test and reshape theories by further experimentation. As-filed patent specifications are now usually published after about 18 months from the priority application date.

Typically the term of a patent is 20 years from the date of application for it (if all fees are paid) and the rights granted to the patentee are to prevent others without his consent making, importing, using and disposing of in the course of trade the things or methods claimed, or the products produced from

INTRODUCTION

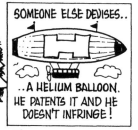

the claimed methods. It is a monopoly; proof that the infringer followed the patentee's teaching or copied his products is not required. Asserting patents is in most jurisdictions solely a matter of private civil law, not public civil law or criminal law.

A patent for an invention does not in most countries give the patentee the absolute right to work his invention himself; he must respect earlier 'master' patents of others for inventions of which his is an improvement but not an alternative outside their scopes. Some jurisdictions provide for compulsory licences in situations where voluntary licences are denied to worthy improvements and where patents are not adequately worked or demand for the patented product is met by importation.

Patent laws and procedures (including important international conventions) are complex and the protection properly due to an inventor can be denied through inexpert handling from the outset and a lack of competent professional advice. There is a published list of qualified patent agents and attorneys in most industrialised countries. Patenting is quite expensive as the chart on the facing page indicates.

DESIGNS
In business, commerce and industry we use the word 'design' to signify a variety of different qualities. We talk of the design of a simple printed form, a spreadsheet, a book-cover, a fabric, an exhibition stand, a corporate logo, a suite of furniture, a building, a stage set, a production plant, a gearbox, a car, a boat, a light fixture, a spectrometer, a can-opener, a printed circuit, a lock, and so on. In some cases visual impact is the sole concern; in others visual appeal is important but utility for its intended purpose is a key element of 'the design'; and in some cases the working interrelationship of parts, the structural arrangement, and the functional interactions are all that matter. Generally, the purchasing public and the professional instincts of the designer favour a combination of elegance of appearance and efficient functioning, ease of use, operation and maintenance.

All such designs are the product of creative effort by individuals or teams and much skill, flair and expertise may have been involved. They deserve protection against unauthorised copying and in all cases the law provides avenues for protection. This law is a compendium of laws reflecting the diversity of 'design'.

INDICATIVE PATENTING COSTS (£'s)

Country	Filing costs and fees to grant†	Indicative agent's charges for prosecution to grant	Maintenance/renewal fees (years)		
			0 – 5	0 – 10	0 – 20
UK alone	200 plus, say*, 500	say, 500	80	650	3,000
CWE via European office (5 countries)	4,200 plus, say, 700	say, 500	500	2,000	13,000
USA	350	800	250	750	1,800
Japan	1,000	600	160	600	1,400
Brazil	950	300	120	500	1,800
Norway (additional to UK/CWE)	900	1,200	200	700	6,200

† Including agent's charges for handling, drafting and translation

* N.B. The potentially large variable in the charges for getting a patent are:
(i) the agent's charges for drafting the initial specification which are, of course, related to complexity and length (and prior art searching charges are NOT included above),
(ii) agent's charges for prosecution (related to complexity and relevance of office prior art citations) and
(iii) translation costs (again related to length of specification and, to some extent, complexity of field).

Some are freestanding, as are patent law and trade mark law described elsewhere, and some are overlapping as in the UK where there is Design Copyright Law, Registered Design Law and a new Design Right Law. These last three will now be considered in the context specifically of industrially applied designs of articles, whether purely decorative (eg an ornament or piece of jewellery) or purely functional (eg a gear box) or elegantly useful (eg a coffee percolator).

I will concern myself only with UK law but it is a reasonably safe assumption to say that the principal industrialised nations have similar or equivalent laws to UK Registered Design Law and to UK Copyright Law for protecting designs that are distinctive. (In the USA, design law has evolved down a somewhat different route from the copyright origins of UK law. The USA grants Design Patents for new and non-obvious ornamental designs of utility articles as it grants Utility Patents for new and non-obvious inventions, all under US Code, Title 35.) In the UK, the main problem areas are unregistrable designs and rights deriving from copyright in design drawings which, even if new and original works, do not, when applied to produce articles to that design, result in three-dimensional artistic works as understood by UK Copyright Law. A further special category are must fit/must match parts which have received both judicial and statutory attention.

APPLIED DESIGN

(a) REGISTERED DESIGNS
This is a monopoly of the right to apply a particular new design to articles of the class for which it is registered, granted on application.

Here, design means features of form, shape, ornament or pattern applied to articles and sets of articles. It is not concerned with the function an article may have to perform nor with its principles of construction and operation which are the province of patent law. Design is therefore judged by appeal to the eye not the intellect. The eye is an instructed one able to recognise and ignore immaterial and common trade variants but protection does not rely upon any concept of good or bad design according to whimsical notions of fashion or taste. Purely functional designs are disqualified as are must fit/must match designs.

Additionally it is a requirement (yet to be clarified by case law) that the aesthetic appeal of the design must be important to purchasers and users. (See Appendix I for further information).

(b) (UNREGISTERED) DESIGN RIGHT
This is a new automatic property right which is targeted at, but not restricted to, non-aesthetic industrially applied designs. Design Right can co-exist with Registered Design protection for the same design of article. Like Registered Design protection it does not apply to principles of construction or to must fit/must match designs. The particular features of the new Design Right that set

it apart from Registered Design protection and give it greater relevance to the mainstream of technology protection and licensing are set out in Appendix I.

These new and revised design laws are conceptually a distinct advance from the previous confusing, indeed quirky, legal protection given to industrially applied designs. They are not the whole story since some aspects of Copyright law continue to apply, as briefly discussed below.

What particular problems arise as a result of the new Design Right law and the amendments made to Registered Design law?

Principally: interpretation of relevant aesthetic qualities (Registered Design law), deciding the scope of the must fit/must match exclusions (Registered Design law, new Design Right law), and proving a design was made before the new Design Right law came into force (and so is excluded from it) or that under the special qualification rules a new design does not enjoy Design Right protection because the originator or marketer is not a 'qualified' person.

Intellectual property laws have always provided relief from damages for innocent infringers. It is noteworthy that the new Design Right law empowers the Court in determining damages for infringement that is not innocent to take into account the flagrancy of the infringement.

COPYRIGHT

(a) COPYRIGHT IN THREE-DIMENSIONAL 'ARTISTIC WORKS'
Artistic works such as works of architecture and works of artistic craftsmanship have always enjoyed copyright protection in their own right and they continue to do so for a reduced term if they are replicated industrially by or with the consent of the copyright owner. Decided cases suggest that the standard to be met if an article is to qualify as a work of artistic craftsmanship is quite exacting so that creators of these sorts of articles that are intended to be mass-marketed should seriously consider going down the parallel Registered Design route. The protection available by the automatic copyright route is, however, for qualifying works broadly similar to that available by the Registered Design route.

The original drawings for these sorts of artistic work are also copyright. Curiously, if the articles in question qualify for Design Right, actions for infringement must nevertheless be pursued down the available copyright route. The court-made rule that must fit/must match parts are deemed licensed to the spare parts industry is an interesting example of fundamental rights being modified other than by statute. It will be interesting so see if this approach is extended to patented items!

(b) COPYRIGHT IN DESIGN DRAWINGS AND DESCRIPTIONS

It is not now an infringement of the copyright in drawings or other documents recording a new design of article which is not itself an artistic work (see above) to make an article to that design by using the documents or by reverse engineering. The other available avenues for protection must be relied upon, eg Design Right. Old designs still retain for up to 10 years more a restricted form of copyright protection deriving from continuing copyright in their original drawings, and this would seem to prevent reverse engineering of articles despite their not being artistic works. In this case, as with the new Design Right, licences of right are available during the last 5 years of available protection. Computer-generated designs and computer-aided designs are equally copyright.

(c) LITERARY COPYRIGHT

Written documents enjoy long periods of literary copyright and unauthorised copying can be prevented. However, the principal protection for confidential and secret licensed technology embodied in documents is that available under the laws of confidence and through contractual secrecy in the licence agreement.

Nevertheless, documents displaying specific correlated information about a product or piece of equipment which could only with a lot of effort be lawfully obtained by competitors by measurement and analysis are valuable items and copyright protection has its part to play to prevent distribution of copies of those documents.

(d) SOFTWARE COPYRIGHT

UK law has substantially clarified and reinforced the copyright protection available for computer software. Software is now protected as if it were a literary work. Infringing reproduction of the program includes converting it to another computer language or code otherwise than incidentally in the course of running the program. It is important to bear in mind that willful or reckless copyright infringement is also a criminal offence; the maximum penalties are severe.

REGISTERED AND UNREGISTERED TRADE MARKS

A trade mark may be an invented word, a device, a logo, a picture, a name written in a distinctive way, or a symbol. It is applied to goods or used in close physical proximity to goods (eg on the container, bottle or packaging) and it serves to indicate a connection in the course of trade between those goods and the owner of the trade mark — sometimes the manufacturer but sometimes the supplier or distributor. It is not an alternative name for goods of the kind in question and

INTRODUCTION

A TRADE MARK MAY BE AN INVENTED WORD

should never be allowed to become so, since all proprietary rights in the mark are then lost.

Marks are always used adjectivally or as brand names. The goodwill in trade marks that every household knows is enormous since the evident success of the particular goods (eg quality, reliability) reflects on the trade mark and people take that mark to indicate value for money even when modified goods or quite different goods are offered. People do not need to know who owns the mark, and often do not know unless it is also incorporated into the name of the proprietor. In the UK, for example, trade marks that can satisfy the requirements for registration can enjoy the benefits afforded by the Trade Marks Act.

Unregistered trade marks become part of the trading style of the business and can be a significant feature of that trading style. Once goods bearing that mark have acquired a reputation in the market place anyone who unfairly seeks to trade on that reputation by using the same, or a deceptively similar, trading style for his goods is committing the civil wrong of 'passing off' and can be sued. Mere use of the same mark (if it is a dominant feature of that trading style) on the same sorts of goods can amount to passing off. It is a question of reputation in the mark.

Registration of a trade mark under the Trade Mark Act, even one that has been widely and long used, gives advantages. Prior use is a help, not a bar, although for an inherently distinctive mark an intention to use is all that is necessary for registration (assuming of course it is not confusingly similar to other marks registered for the same kind of goods or for goods of the same general class). An inherently distinctive mark would ordinarily be registered in Part A of the register for a designated class or classes of goods and immediately entitles the proprietor to stop others using that mark on those goods. For a registered mark, there is no need for the proprietor to prove reputation or to show

11

that the public are not confused by its similarity to other marks. A trade mark that is not inherently distinctive but is capable through use of becoming distinctive can only be registered in Part B of the register but gives similar monopoly rights unless the alleged infringer can prove that the public are not confused by his use of a similar mark or would not assume any connection between his goods and the registered proprietor of the trade mark.

The term of protection for registered trade marks is perpetual, if due fees are paid. Rights to prevent passing off (and therefore, to some extent, in unregistered trade marks) are also perpetual.

SUMMARIES

PATENTS
(1) A statutory monopoly for creativity in applied science and engineering. Patents for new products and industrially applicable techniques/processes are granted, country by country, on application. A block European Application designating chosen countries is possible. The Patent Cooperation Treaty (PCT) also offers a valuable procedural option for multi-national protection.
(2) The term of monopoly is 20 years from the application date (if fees paid) for patents in European countries but 17 years from grant of patent in the United States of America.
(3) There must be an invention step (ie not obvious) over 'prior art'.
(4) 'Prior art' means all public knowledge at date of application for patent (not at date of making invention). US rules still treat invention date as controlling if patenting is not too long delayed.
(5) Own non-confidential disclosure will destroy patentability, if before date of priority application, except for USA patents for USA invention if applied for within one year of first disclosure or commercial public use.
(6) A patent is infringed by working within the claimed monopoly, including importation of product made abroad by a process covered by the patent. Relief includes an injunction, damages or an account of profits.
(7) A patent does not give the patentee the right to work his invention. Other local patents of third parties could prevent working.
(8) A patentee does not (unless local statutory conditions on inexcusable non-working/inadequate working are met) have to license his patent.
(9) Secret commercial use by a third party gives him a personal right to continue use in the EEC and in some countries will even invalidate the patent.

In the USA, if the prior use is not such as to defeat the patent, the prior user has no rights under the patent.
(10) Professional advisers: Patent agents/attorneys.
(11) Patenting is expensive — say up to £2000 per country for 10 years cover. (Includes agent's fees, translations, official fees.)

DESIGNS
(1) A monopoly for aesthetic designs applied to articles of manufacture granted, country by country, on application.
(2) The term of the monopoly is 25 years in the UK (if fees are paid). An inexpensive form of projection.
(3) To be registrable, the design must be new judged by its appeal to the eye. Function is irrelevant and purely functional designs are not registrable.
(4) Newness (whether in form, shape, ornament or pattern) at date of application is essential.
(5) Rights are infringed by making or producing articles to that design (or moulds or patterns) without consent. No requirement to show copying. Relief is as for patents.
(6) Design copyright overlaps for articles that are artistic works and gives similar automatic rights in most cases, without formality, if there has been direct or indirect copying of the design.
(7) Additionally, design right protection is now given for purely functional designs and designs devoid of aesthetic appeal. Must fit/must match designs excluded. Design Right period of protection is 10 years from date applied industrially. Design must be original but need not be new.
(8) Professional Advisors: Patent Agents/Attorneys.

REGISTERED TRADE MARKS
(1) Effectively a perpetual national monopoly (if fees paid), but an honest concurrent user is protected.
(2) A registrable trade mark is any invented word, any pictorial device, symbol or logo, any name written in a distinctive way (eg a signature) or any other distinctive mark.
(3) Inherently distinctive marks can be registered in the UK in part A of register. Marks that through use are capable of acquiring distinctiveness are registrable in part B, usually after extensive use.
(4) Prior use — no bar. Indeed helpful.

(5) Purpose of trade mark to indicate a connection in the course of trade between particular goods and the owner of the trade mark.
(6) Registration is for classes of goods.
(7) Use of the mark in the UK by others without consent, if used as a trade mark on goods within the class or classes for which it is registered, is an infringement actionable by the owner. But no relief for part B mark if public not confused and would not assume any connection with owner of registered trade mark.
(8) Not too expensive to get.
(9) Professional advisors: Trade mark agents, and specialising patent agents/attorneys.

None of these statutory intellectual property rights directly gives protection for confidential, secret and proprietary technical knowledge either against prejudicial disclosure or publication (since copyright concerns form of expression not intellectual content) or against prejudicial use (except to the extent a field of use is dominated by patents still in force). Were there not such protection available, or were it of application only to higher orders of information such as uniquely held trade secrets, licensing would not be the major international activity that it is. But all sorts of technical knowledge are protectable, if due precautions are taken.

THE PROTECTION OF CONFIDENTIAL INFORMATION

When information is acquired as a result of confidential dealings, or through the performance of a contract, and it was implicit in the relationship or expressly or implicitly agreed that the information was supplied only for a specific purpose and was not to be used for other purposes and was to be held in confidence, legal obligations are created not to disclose or use the information except in approved ways and these will be enforceable under contract law or, as the case may be, through an action in tort (civil wrong) where there has been prejudice to the supplier of the information.

This is the legal basis for the protection of a licensor's confidential and secret information. The protection rests solely on the true construction of the licensee's self-restricting obligations, normally as determined by express contractual undertakings but sometimes by a necessary implication the law reads into the relationship and the nature of the transaction. With patents and other forms of legally created monopolies the precise opposite applies.

Within the proper scope of the monopoly given by the patent or the registered design, the licensee has no rights of use except those granted by the licensor or those conferred by rule of law, for example under the EEC 'exhaustion of rights' principle (see below), or consequent on the purchase of goods marketed by the owner of the monopoly or by others with his consent. And so, in the context of licensing, an unpatented use of licensed information is unrestricted except as 'agreed' otherwise by the receiver and for patented information use continues to be preventable by asserting the patent except to the extent permitted by a licence from the patent owner.

Contractual undertakings in licence agreements are not, of course, exempt from the impact of the relevant proper law of contract as to the validity and enforceability of contract terms nor are restrictions imposed (or restrictive effects of terms) immune to challenge under, for example, the patent laws, competition laws ('anti-trust') and laws concerned with restrictive and unfair trade practices. But, realistically drafted terms that clearly address genuine ills and are reasonable in their impact (which is all they need to be to protect the licensor's legitimate interests) have nothing to fear from these legal proscriptions.

When the purpose of the licence is to instruct another party (the licensee) in the practise of a particular art, patent considerations — at least those relating to the licensors' patents — may be very much secondary issues. The licensor may have chosen not to seek comprehensive patent cover in view of the practical difficulty of policing competitor's operations behind closed doors, or of proving from a product that a patented process has been used. The licensor may have decided that the publication of inventions and useful information that any patenting entails was too great a sacrifice of valuable knowledge. He will have accepted the risk that others might patent the same technology and be content to rely on statutory exemptions allowing his commercial activities to continue, where available.

Alternatively, he may have decided that strong, broadly-based patents were not obtainable because of what is already in the public domain, even though the significance of that information may not generally be appreciated. It is a common experience of patent specialists that the 'average man skilled in the art', whose knowledge is the legal base line against which patentability is assessed, is frequently judged to be much more knowledgeable of the prior art and much more alert to the practical and commercial significance of information buried in a mass of published literature than ever are those who actually practise in a given field. It may be, of course, that the technology is mature and basic patents have already expired.

Similarly, the territorial interests of the licensor in the early years of making the basic discoveries and developing the technology may have been limited to his home territory and his immediate manufacturing and export territories so that broad geographical patent cover may have seemed an expensive and unnecessary luxury (see the chart on page 7). Whatever the reasons why the licensor does not have much in the way of patents covering facets of the technology being transferred, it is nevertheless customary, and indeed prudent, for the licences to include rights under patents (if any) of the licensor covering the relevant field of production or product in the appropriate territories. This is, as a minimum, a fail safe provision, because licences not granted are licences withheld and the licensee may not be told or be able to ascertain definitively from public records which of the licensor's patents are and are not relevant to the licence he is negotiating and accepting. Additionally, if suitably framed, the patent licence clauses anticipate any patenting the licensor may yet decide to seek on secret aspects of the technology, as well as anticipating possible developments of the technology by the licensee in directions covered by patents of the licensor that at the time of the technology transfer were thought to relate to non-commercial or uneconomical options.

Finally, it must be kept in mind that patents and rights in them are a legal creation whose scope and nature are determined, country to country, by national laws and whose exercise will usually also be regulated by national laws and international treaties. One must always, therefore, construe the patent licence on the basis equally of the expression of that licence in the agreement and of the laws which create those patent rights and regulate their exercise. By contrast, rights to use confidential information will be essentially solely a matter determined according to contract law and, at least in developed countries, there is mercifully considerable uniformity of contract

INTELLECTUAL PROPERTY IS THE BASIS OF YOUR COMPETITIVE EDGE, NOW AND LATER. GUARD IT WELL!

laws as applied to licensing and considerable freedom allowed to parties to contract on whatever terms they choose. In controlled economies and developing countries, technology transfers are centrally controlled. Unapproved agreements may be illegal, or void, or unenforceable and approvals for technology transfers that do not square with national policies and aspirations will be at best difficult and at worst impossible to secure.

2. THE BUSINESS IMPLICATIONS

Licensing is trading in technology and, as with trading in goods, there is the perspective of the seller and the perspective of the buyer. The motivations and pressures that lead organisations to offer or seek licences, and the consequential effects of doing so, are discussed in this chapter. Licensing touches fundamentally on business policy and technology management, including R&D programmes. There are potentially large 'pluses' and large 'minuses'. Aside from the immediate and short term effects of a proposed licence deal which probably are the justification for the bargain which is struck, the longer term implications can be considerable for licensor and licensee in their different ways. The general corporate policy on 'licensing-out' technology cannot sensibly be separated from the responsibility for overall management of corporate technologies, even though the day-to-day responsibility for licensing can, if the volume of demand justifies it, be delegated to a licensing executive. Whether you license, what you

'LICENSING IS TRADING IN TECHNOLOGY'

license, where you license, and whom you license, as well as to some extent how you license, are all business decisions needing continual review. At its baldest, what is the point of licensing today your core technologies on which your mainstream business of making and selling products depends, if tomorrow your licensees oust you from that business?

Licensing in available technology such as a new or improved manufacturing process can indeed be the ideal solution to today's problem of a deficient technology base — and it will usually be cheaper than any feasible alternative and quicker to accomplish. The immediate balance of opportunity versus cost may be most attractive, but, as is discussed later in this chapter, the implications and effects have to be seen on a broader canvas.

THE MOTIVATION TO SEEK OR OFFER LICENCES

For many organisations who possess saleable technology there is no doubt that the principal factor that leads them to seek to license that technology is the opportunity of earning revenue in the form of fees and royalties, and a lot of licensing is done for this reason alone. Many organisations exist to develop or perfect technologies and then license them. Their business is technology development for sale and some have an impressive track-record and are prominent in the fields of technology in which they have chosen to specialise. Their life-blood is in part effective R&D programmes and pilot plant studies and in part the concentration of the experience of their licensees which they achieve through the contractual force of their licence terms which ensure feedback of developments and operating experience. State-financed research establishments and universities are sources of technology and innovations which are licensed to earn money even though perhaps not generated for that purpose.

The main international engineering contractors are also licensors of their own and others' technology as a vehicle for their main business which is selling engineering man-hours and plant hardware. All these bodies are, in the main, indifferent to the market consequences of proliferating competition in products, of over-capacity, and of disturbance of the relative competitiveness of given companies.

In contrast, operating companies whose main business is the making and selling of products are far from indifferent to these market consequences; yet they are major suppliers of technology for cash. They will have reached their decision to run a side-show business of selling technology after first concluding that their mainstream activities and aspirations will not be undermined or seriously threatened despite relinquishing exclusive possession of valuable information or rights and equipping others to be more effective competitors. What they might have done instead is considered below, as are possible beneficial consequences of licensing in addition to revenue earning.

Licensing can occupy the centre stage of a business development strategy. A simple case would be an arms-length exchange of patent rights in the same field or different fields. The reason may be to avoid actual or threatened patent conflict or to remove the risk of infringements by lifting constraints on existing operations or R&D programmes.

Another example would be licensing a company based in a territory or market sector presently not served by the licensor with the aim of using the licensee to establish market acceptance and penetration in the same sorts and quality of goods as the licensor produces. In this way, penetration of the licensor's products into that territory or market sector will be easier (subject of course, if goods have to cross national boundaries, to tariff barriers and other measures to protect local producers as well as any limited period of exclusivity which may have been conceded as an inducement to the licensee). Eventually the licensee may become an attractive acquisition target. Such a strategy as this will, however, only be adopted after other forms of business development have been considered and rejected.

It is an oversimplification, but in no way misleading, to say that licensing technology stands lower in the hierarchy of ways of exploiting technological strengths than manufacture of products and sales.

Suppose an unsatisfied demand exists for products of the same sort as a company produces. It may be a new product lacking close competitive equivalents or, at the other end of the scale, it may be a commodity that can increase its market share because it can be profitably offered at a significantly lower price as a result of a more economic production technology. The company will ordinarily first consider increasing its production and sales. If the product is a pillar of its business, the unsatisfied market is in its home territory where production facilities can be readily expanded or is in its backyard where it can readily sell, and if there is no risk of the company overreaching itself by the required growth then expansion must be a preferred course of action. But things are not always, or often, like that. The freight costs of sales to distant markets, tariffs and other central Government measures designed to protect local producers against imports, and lack of reputation or intimate knowledge of the market, all contribute to a would-be exporter's costs and business risks.

Another possible course would be to establish a manufacturing and selling subsidiary or joint venture in the territory where the market opportunity exists. The rewards then will be dividend on ownership interest (which you hope you will be able to expatriate to an acceptable extent) with the possibility, at least in early years, of profits on 'seed' consignments of finished products, supplies of part-finished items and sub-assemblies and supplies of required raw materials, components or services (especially if they can enjoy preferential tariff treatment) as well as licence fees. Many countries place limits on the proportion of equity control held by foreign entities either in basic industries or across the board to varying degrees. If that course also is rejected, then consideration turns to arms' length licensing to unconnected companies for cash revenue.

However, the supply of technology to subsidiaries, affiliates or other forms of joint venture is equally licensing and needs to be approached conceptually as such. Indeed, if the venture is in a foreign country (and even if the licensor is the sole owner) the transfer will not merely be conceptually a licence but will have much of the form and incidents of licensing between strangers at arms' length.

The following are relevant factors:

(1) Technology and patents rights should be rewarded by fees and royalties. Dividends reward ownership interest. A failure to charge for technology and patent rights is tantamount to making a gift to minority shareholders and joint venture partners. It is denying the licensor's owners a right to dividends because he is subsidising his licensee. It must be realised, too, that a licensor's ownership interest in a licensee may be voluntarily or compulsorily relinquished with or without adequate compensation. The licensor, if he is a part owner, may acquire extra shares in lieu of expatriating fees and royalties, but this is equally a reward for the technology supply.

Additionally it must be appreciated that, for corporation tax purposes, the supply of technology is treated as an export of an asset by many countries, including the UK and the USA and a gift or sale at an unrealistic price will leave the supplier liable to tax as if a proper price had been charged.

(2) Payments of royalties and licence fees may require approvals in the payor's country before they can be remitted and they may be subject to tax withholdings. Similarly payment of fees for services rendered in the licensor's and licensee's countries may be subject to approvals and substantial withholding taxes. The exercise of patent rights may give rise to VAT payments on royalties or deemed royalties. Thus the licence agreements need to deal with approvals, payments of

tax, and Double Taxation Treaty provisions as meticulously as if the licensor and licensee were unrelated companies.

(3) The mere acquisition of foreign technology by a company resident in many controlled economies and developing countries is a matter needing official approval both as to whether the technology should be acquired at all and as to the terms on which it is to be acquired. The supply of services and equipment from abroad is commonly equally scrutinised to ensure that nothing is being bought in that cannot be supplied locally. It may be a condition of investment grants (or other fiscal inducements) that a high proportion of engineering services or plant items be procured locally.

(4) Export of technology is a matter of national defence, strategic, or even political concern. Licensees of technology of US origin are for example required to agree to observe the Export Control Regulations of the US Department of Commerce which are directed at supply to, for example, Iron Curtain and Bamboo Curtain countries of technical data of US origin and the 'direct products' of it.

(5) Both parties will be concerned to meet statutory requirements on the grant of rights in intellectual property such as patents and trade-marks so that rights are properly vested, acquired rights can be registered, and the integrity of the intellectual property is not prejudiced.

(6) The parties will desire, as a matter of good discipline and definition, as well as to accommodate and respect the interests of minority shareholders, to set down precisely the scope and nature of the performance obligations being assumed and the responsibilities and liabilities of the parties in the event of, for example, loss or damage to property, injury or death caused to individuals, including products liability. In particular, the licensor will wish to ensure that failure, inadequacy or even negligence in his performance will not expose him to liabilities to third parties by attribution of a relationship of principal and agent, or partnership, between him and his subsidiary or joint venture. Where liabilities cannot be avoided, he will want indemnities and/or appropriate insurance. Similarly, he will not want to be liable to his licensee for costs, expenses, losses or damages which the licensee incurs as a result of using the transferred technology; he will seek waivers of liability beyond the conventional limits of performance guarantees.

(7) Under Article 85 of the Treaty of Rome, agreements that may affect trade between Member States and have as their object or effect the prevention, restriction or distortion of competition within the Common Market are prima

facie bad, and need for validation and enforcement (as well as the avoidance of fines) approval by the European Commission. Outside the 'single corporate entity' concept, joint ventures can give rise to concerns. Acquisitions and mergers are liable to scrutiny and objection under national mergers and monopolies laws as well as under the 'abuse of dominant position' provisions of Article 86 of the Rome Treaty. US antitrust laws and German anti-cartel laws are well developed and are powerful moderators of intercorporate arrangements. National laws concerned with restrictive and unfair trade and business practices are never very far away. The collective purpose of all these laws is to outlaw arrangements which have certain objects or effects (or impose certain types of restrictions) which are considered indefensible in any circumstances or are judged on balance to be excessive or against the public interest.

A detailed review of European Competitive law, and its maturer antecedent US Anti-Trust law, is beyond the scope of this work. These laws are not, however, antagonistic to the mainstream of technology licensing. They recognise that technology licensing within the framework of intellectual property law enables innovation to be both encouraged and rewarded and, in the case of confidential secret and proprietary technologies, encourages the wider availability of improved products and improved production processes which, aside from immediate economic benefits, creates a multi-centred opportunity and incentive for yet further improvements to be made. Some critical comments on the way Article 85 of the Rome Treaty has been implemented are included in Appendix I.

A licensing programme can be supplemented by leasing. If, for example, a company possesses a proprietary or patented process of making goods, whether articles or chemicals, and possesses already a fully developed production tool or piece of equipment or catalyst which is especially useful for commercially exercising the production process, that company may license third parties under process patents or in respect of the process technology for a royalty on production output and couple the licence with an offer of a lease for the perfected production tool, equipment or catalyst against the payment of lease fees. The licensee is not obliged to take a lease and the patent licence or technology licence will be unaffected by his choosing to use other means available to him from other sources. However, if the licensee does not in fact have alternative but equally effective means available, say because it does not exist in such a developed or proven state or because the licensor also holds dominant patents on the best versions, the licensee will be attracted to the lease.

The licensor/lessor then operates a lease business alongside the licence business. The advantage to a lessor is that, by leasing, the means of production passes but not the key knowledge of how to create that means. By retention of ownership, he can adopt measures and extract undertakings that effectively prevent key secret and confidential information being discoverable from the leased item or being passed to others. In most cases, assuming the lessee has security of supply and, if appropriate, a services contract, the lessee will not be unhappy; he can run his business because he possesses all the necessary effective means and all the necessary rights.

Earlier it was said that licensing creates a potentially more effective competitor, and yet paradoxically licensing can stifle existing competitiveness. Someone once suggested that the best way to put a stranglehold on your competitors is to tell them your secrets in confidence. Like all exaggerations it has a germ of truth. It applies at the assessment and evaluation stages that precede the actual grant of a licence as well as after the license is granted.

At first sight, it does seem to be a contradiction in terms to say that licensing someone will limit his freedom of action. Nevertheless, even for arms' length patent licensing which, on the face of it, does no more than simply confer a freedom from patent suit, there may be consequential effects which reduce a licensee's competitive potential. In every licensing situation the essence of the matter is, "What would the prospective licensee otherwise do?" Consider the interaction between a manufacturing and selling company (the potential licensor) and another company (the possible licensee) which is determined to get into or become larger or more economic in the same field of manufacture and selling. If our possible licensee does not already have in-house 'off the shelf' proven technology that would be economically competitive in a new plant (and many companies in an industry will not, even though existing operations are fully competitive) that company has a choice. It must either develop the needed technology or acquire it. Development of technology implies an uncertain timescale, uncertain costs, and an uncertain outcome. The technology may not achieve the required criteria, the economic goal posts may move in the intervening period, the market opportunity may have disappeared by the time the development is complete, the learning process involved in any new technology will impose economic penalties, the costs of development may sink a single plant business, the ultimate product from the newly developed technology may meet customer resistance because it is different from what they are used to, there may be a patent minefield, and so on. By contrast, acquisition of technology from a market-leader licensor holds the promise of technical certainty, a definite investment project programme, ascertainable capital and operating costs and a product that the market knows. In addition there will be greatly reduced or non-existent patent infringement concerns. By offering a licence on attractive terms, our potential licensor will find a receptive ear and can hope, by licensing:

- to deter the competitor from developing alternative, and perhaps better, technology;
- stop him turning to competitive technology offered by others so that his manufacturing cost base cannot be better than the licensor's own (usually it will be worse, eg new capital, 'learning costs', scale differences, 'conservative' technology and designs supplied);
- by supply of a complete design package and not the knowledge needed for the design of plants, to make it difficult for the licensee (and reduce his incentive) to become equally proficient in technological terms;
- by legitimate technology exchange of future improvements and developments, to ensure that the licensee cannot outrun the licensor or acquire patents which could provide detrimental leverage;
- to achieve, in appropriate cases, market developments at the risk and expense of the licensee which the licensor can hope to benefit from;
- by a patent licence, if the licensee is a significant force in the industry, cause other competitors to be less confident that the patent is invalid or easily evadable.

    The impact of licensing on the pace and direction of ongoing research and development in the relevant field of technology and products needs to be considered. The development of new and radically different technology often calls for a major investment in equipment, facilities and people. It is a high risk activity. Many companies simply cannot afford to do it and consciously decide to rely on purchase for their technology needs. Some companies are prepared to undertake the risk and expense but justify doing so in part by the reward they expect to get from licensing. Commonly, the costs of development are increased because, if licensing is intended, the technology base will have to accommodate varying industry conditions, eg different raw materials and feedstocks, different conditions for utilities, different desired product specifications, different climates and different scales of production. A technology is only as versatile as its most constricting essential part. Additionally, the developer of a technology will be prepared to accept technological risks and uncertainties, by backing the skill and judgement of its specialists and field operators, to a degree that would be unacceptable to a licensee. And so, a higher standard of technical proof and confidence will be required if the technology is to be generally licensable.

But, of course, once a new technology is developed to the point of first commercial use and is then licensed to others, the technology does not stagnate. Developments are made, and there is a growth of experience as new situations are met in the course of operations. Many licence agreements call for exchange of improvements and experience between licensor and licensees, at least for a period of years, and this includes indirect exchange between licensees using the licensor as a central collator and clearing house. These arrangements work poorly or well depending upon the commitment to them of all parties. Some people would argue that a grant-back obligation is a disincentive to do much process improvement. Others would say that access to a broad base of others' improvements and experience is a good reward for contributing the results of one's own development work which one would probably want to do for one's own benefit anyway.

Licensors want feedback from their licensees because they then ensure that their technology in their hands stays ahead and that they always have the most up-to-date technology for licensing to others. The prospects of future licence revenue are enhanced and, if the licensor is also an operator, he has operating benefits as well.

THE ADVANTAGES AND DISADVANTAGES TO LICENSORS AND LICENSEES

At this point a general note of caution should be introduced. Licensing is a responsible activity. When licensing a subsidiary or joint venture, the commitment to the success of the technology transfer and of the resulting operations themselves may be assumed. Agreements apart, what needs doing will be done; planning, design work and studies, process engineering and detailed engineering, training, start-up services, market development, raw materials and components supply, troubleshooting. When licensing third parties at arms' length the commitment is different and those who are responsible for managing the licensing function need to be aware of what the implications of this are. Here are a few points to consider:

(1) Your technology reflects the infrastructure of your operations; you cannot transfer that to your licensee's location. Consider, therefore, the business environment, the social environment and the company environment into which the technology is going. Consider whether you should link up with an experienced

'PROPER PLANNING PREVENTS POOR PERFORMANCE'

international contractor who has already done business in the licensee's territory and who may be better placed than you to identify, recommend and perhaps procure solutions to problems. Do you offer capital-intensive or labour-intensive variants of unit operations? Batch, semi-continuous or continuous versions? How important is operator skill? What need for 'show-how', training, operator/supervisor secondment? Indeed, how sure are you that you really can design and supply something other than a replica — a Chinese copy — of what you actually do?

(2) In licensing, success breeds success. A disgruntled licensee, even one who had undersold his needs and oversold his competence, is bad publicity.

(3) Will the licensing function get the support it needs from the research/technical/production functions who exist primarily to service in-house business? Will your needs be low on their priorities? Will they field their second eleven?

    To some extent, the perspectives of a licensor and licensee are opposite sides of the same coin but there are factors that merit consideration from the standpoint of the licensee or prospective licensee. The role of 'licensing in' technology or patents within a business development strategy is an important topic.

    Suppose a company has identified a new line of business, or manufacturing activity, which offers good prospects of a significant return on investment.

It might be a complete diversification which makes a good fit with its management and technical skills, or be a backwards or forward integration promising better roll-through economics and more added value. Acquisition of an existing operator who has good technology and assets on the ground, or purchase of the relevant business, technology and production facilities from that operator might be a first consideration. Issues such as its availability on the right terms, the location and logistics of its operations, the practicality of severing a business even if offered, incompatible corporate cultures and organisations and unwanted alien product lines, the liabilities and commitments of the prey company all come to the fore and may show this strategy to be impractical or undesirable. Licensing in technology, if available of course, could be the better route and preferred over the only other alternative of in-house development in conjunction with purchase of any necessary patent licences.

FINANCIAL FACTORS
On purely actuarial grounds, redevelopment of technology that already exists and is available for licensing is a needless expense. In general, development costs far outstrip licence fees and royalties likely to be payable. Japan and Germany have historically been large net importers of technology. The UK has had a sizable trade in both directions. The USA has been a major net exporter of technology. This implies a general receptiveness to technology transfer which must have an economic justification.

STRATEGIC FACTORS
But another factor favours 'licensing in' over in-house development and that is time of availability. Developed technology will be available for use, if at all, much later (3-8 years?); licensed technology is available now.

It must be stated, however, that a technology owner cannot be forced to reveal its secrets to another; it is not obliged to license its technology. Patent licences on the other hand tend to be more readily conceded. There are special influences and attitude conditioners at play here. The patent holder may fear that refusal will lead to unlicensed use in secret, or to a vigorous assault on the validity and scope of his patent. He may consider himself vulnerable to a compulsory licence order (which most national patent systems provide for in defined circumstances) and so reason that he would do better from a voluntarily negotiated licence. He might comfort himself that since only patent licences are being granted and not rights to use his secret and confidential technology his competitiveness will be little affected. He may simply say to himself that one day the tables may be turned and he may find himself needing a patent licence; intransigence today without sound commercial reasons might store up difficulties later. There is no doubt that the willingness of companies in an industry to offer their technologies or patents for licensing is influenced by how they perceive the willingness of others to reciprocate.

In addition to its role in business development by complete diversification or a horizontal or vertical integration of new businesses, licensing in is directly relevant also to existing businesses. In particular, it can enable a company to re-establish competitive operations at the same or a different scale.

All manufacturing facilities have a practical capacity limit and there is a limit on the extent to which they can be economically enlarged, modified or extended. They have a finite functional life. It is frequently the case that the right type of investment at the date of decision to proceed is rarely the right one to repeat for a replacement towards the end of the useful life of the initial investment nor, commonly, is it the right one for added 'capacity', say, 5 years into the operating life of the initial investment. Changes in the availability and cost of raw materials and of energy, scale effects, technology advances, market growth, and more demanding product specifications can all militate against mere duplication.

EFFECTS ON R&D

Most companies have certain core businesses to which R&D effort is devoted but equally most will have neither the resources nor the skills to establish dynamic R&D programmes across the full complement of their business activities so as to be technologically self-sufficient. Indeed even in core businesses the effort may be directed at improving the familiar technology rather than at radically new technologies. Whatever the policy may be, R&D is notoriously uncertain in outcome and the real possibility always exists that other organisations will make the advances that render one's own technology obsolete for a new investment. Thus, technology acquisition through licensing is likely to be needed sooner or later in most businesses if the economics of the total business are to remain competitive. The best one can do is to avoid being over-dependent.

MARKET EFFECTS

In certain market sectors, downstream consumer industries are geared up to use only raw materials that meet certain stringent quality specifications. The polymer and fibres fields, electrical and mechanical components, and the cosmetic and food industries are examples. Additionally, industries that purchase effect materials (ie products sold against a performance specification and not a composition specification) are resistant to new sources of materials because of unpredictability of end performance. Yet again, where new legislation or statutory instruments, or official approvals are a prerequisite to marketing a new product (usually after elaborate and time-consuming laboratory, clinical and field testing) there is understandable reluctance to pioneer novelty. For all such reasons a would-be producer of a product may be constrained in his choice of technology for a new business or a proposed expansion, and acquisition through licensing may be his only realistic option.

PATENT ISSUES

Both technology development and business development can be obstructed by third party patents and acquiring a licence will offer a way through. No need for know-how is implied. The objective is to take a licence under patents that cover what the licensee is otherwise technically equipped to do. The patents may relate to production of the licensee's products or, alternatively, cover fields of important uses of those products. These downstream uses may be in-house operations or operations done by customers in which case the licence must allow for sub-licensing or 'label-licensing' (see later). Patent licences may be needed when important export markets are closed off. Patent licences may also be desired when embarking on a technology development programme. Designing around, or through, patents can be technically inefficient or risky.

Patent applications take a while to reach the status of granted patents and commonly what first emerges as evidence of relevant patent activity is a published application that has not yet been critically examined. As mentioned earlier and in Appendix I, many applications suffer curtailment of scope in the course of examination. Even when granted, a large proportion of patents are vulnerable to revocation for want of inventiveness against the state of the art (which no patent office can know in full or comprehend objectively). There are many difficult questions confronting a manufacturer or developer who sees a patent or patent application standing in his path or lurking close by. For example:
- should he do a comprehensive validity assessment before opening discussions with the patentee/applicant or deciding to ignore the patent?
- can the patent be evaded and at what extra development cost?
- will an early approach give the patentee/applicant the opportunity to remedy

defects in his patent or to re-enforce it by allowable amendments to entrap more tightly what the manufacturer or developer wishes to do?
- will the patentee/applicant, if he is already a manufacturer, be able to take practical steps to crowd out the developer either by his own commercial activities or by licensing the developer's competitors from whom perhaps he has less to fear?
- will an approach for a licence made late in a development programme or sales promotion merely mean the stakes will be higher because it will be plain that there will be much less flexibility to design around a patent and an added urgency to the negotiations?

But suppose a licence is refused or offered on totally unreasonable terms (which is a cosmetic refusal):
- should an enquirer who genuinely doubts the patent's validity seek patent revocation? Costly, time consuming, uncertain and not always possible procedurally. For example, in the USA outside of an infringement suit there is limited scope for challenging an issued patent.
- should he go ahead and 'infringe' if he sincerely believes that the patent is invalid and could not be remedied to cover his operation, and put the onus on the patentee to sue?
- what relief might the patentee get if the patent survived an action for revocation, and was held infringed? National laws vary as do court attitudes since, although some relief may be a right, the form of relief will be decided on equitable principles. However, an injunction and damages are usual consequences.
- will the patentee be induced by the costs and uncertainties of patent litigation to seek an out-of-court settlement? Will he prefer to license, after all? Paradoxically, this could lend standing to his patent in the eyes of others.

Each situation must be assessed on its facts and circumstances in the light of the patent itself, the relevant patent laws and the attitude of the courts of the territories having jurisdiction. It is a complex area demanding specialist advice and sober technical and business judgements.

Provisions in licence agreements under which the licensor and licensee agree to exchange research results in the relevant field or, at least, production improvements and developments that have been implemented commercially can, as suggested earlier, be a useful side-benefit of licensee status, especially if the technology does not relate to a core business. Indeed, the opportunity to have access to a broader base of research and production experience can be a stand-alone justification for a technical information exchange ('cross-licensing') between producers who operate in the same technology field but do not compete in the same markets.

'IN LICENSING, SUCCESS BREEDS SUCCESS!'

3. TECHNOLOGY LISTING AND EVALUATION

PRE-LICENCE SHORTLISTING OF AVAILABLE TECHNOLOGIES

The granting of a licence for production technology is always preceded by an evaluation of relevant technologies available from licensors. This is a phase of activity that must be managed, especially by the potential licensee, with considerable care.

Relevant issues are: the point at which confidential relationships with bidding licensors are created, the contractual form they take, the control of information flow in relation to the purpose for which it is required, the ability to disengage cleanly and without avoidable prejudicial obligations, and ensuring that rejected licensors do not have a case for an action for misuse of their confidential information when the investment proceeds with the chosen licensor's technology.

As a first step, a company that is considering acquisition of technology for a manufacturing venture will make itself aware in general terms of available technologies. It will be specifically interested in technologies used by market leaders and particularly its immediate direct competitors. In many companies this is an ongoing activity of the technical function for areas of manufacture of direct or potential interest.

At an early stage in planning a new venture, the company will seek from known licensors:
(1) Confirmation that they would, in principle, license their technologies.
(2) An indication of their track-record of using and licensing the technology.
(3) Such non-confidential outline economic assessment data and process descriptions as the licensors are willing to supply together with an indication of

the sorts of confidential information they would be willing to provide for closer technical and economic evaluation, plus a copy of their 'normal' Secrecy letter. Some information on licensors' patents might be available at this stage.

(4) Whether and at what stage visits to operating plant, and other licensees, could be arranged.

(5) What, in outline, their 'standard' licence terms are, especially royalties/fees, and whether the licence would be a unit licence or a territorial production licence, and whether there would be export restrictions deriving, say, from non-availability of patent licences in given territories.

At the same time, if the situation allows, an approach might be made to any major manufacturer who uses his own technology and has not yet licensed or sought to license his technology.

The purpose of these initial assessments is to place the enquirer company in the position of being able to draw up a manageable list of those technologies from which a final choice should be made. It is a non-compromising activity, save for betrayal of interest in investment. It does not imply that no in-house technology exists, nor that it is considered inferior.

MANAGING PRE-LICENCE CONFIDENTIAL ASSESSMENTS

Every action taken at this stage is either in circumstances where no relationship of confidence or trust exists or in circumstances of stipulated non-confidentiality, or, very occasionally, confidentiality (non-disclosure) of economic data, which is no great handicap to the assessor company.

The next step involves a more detailed evaluation of shortlisted technologies. This will involve compromising confidential disclosures. The risk of prejudice to the interests of either party is now real, and as stressed above the exercise must be managed and controlled (and not merely by the terms of the agreements) to minimise these risks or to keep them within acceptable bounds.

It is a surprisingly common and dangerous misconception amongst businessmen, technologists and engineers that if disclosures take place without expressly agreeing they are confidential, and without a written 'secrecy letter', there is no actual confidence and no legal impact. This, as explained in Chapter 1, is totally wrong. Detailed evaluation of technologies will often involve the passing of information that is of a kind that is practically useful and not merely (as economic information would be) useful for decision making as to which technology to choose. (But often Licensors do not wish detailed economic comparisons to be broadcast.) The law will not lightly presume a gift of valuable information the onward disclosure of which or the use of which would cause damage to the original source "as the parties must surely, as reasonable businessmen, have appreciated". Licensors will, of course, be at pains to minimise the passing of information that betrays what the technology consists of and how it works, as opposed to what it costs to implement and what it achieves. But, inevitably, a significant proportion of the former kinds of information may need to pass. Detailed evaluation of technologies should always be done on the basis of contractual confidence. There should be a prior agreement setting down the class of information to which the contract relates, and specifying what the non-disclosure obligations and restrictions of use are. Within its ambit, the contract then is the sole determinant of the legal relationship of the parties in respect of its subject matter. It is better to have clear contractual obligations, even if in some respects wider or narrower than necessarily implied by the general laws of confidence, fair dealing and unjust enrichment than to have uncertain obligations.

If confidential discussions do take place before there is a written 'secrecy agreement', then every effort should be made to get a confirmatory agreement drawn-up and signed as soon as possible while memories are fresh and the parties have not changed their positions.

The best way to avoid or minimise prejudice is to avoid disclosing and receiving sensitive and practically useful information. Consider, from a prospective licensee's standpoint, what he actually needs to know for his decision-making process at this stage.

He will want:
- a general description only of what the technology is; raw materials/feedstock types; products, coproducts, biproducts, their condition and quality; types of unit operation; if special items of equipment are needed and for what purpose; manning levels; batch-cycle times, downtime for routine maintenance; whether special catalysts and processing aids are needed and where they are available from;
- an estimated plant cost (stated location, money of the day) — perhaps the cost of main plant items;
- consumptions of raw materials/feedstocks per te/unit product(s); similarly usages of utilities (energy), catalysts and processing aids;
- the cost of a design package and how long it will take to produce it;
- a general patent statement (Licensor's and third parties' patents);

- a fuller account of Licensor's track-record of using and licensing the technology — other licensees, where, when, scale of operations, identity of licensees:
- draft design and licence agreements.

This information enables a prospective licensee to determine what he needs to do to install and operate the technology, and the probable capital and operating costs. It confirms if products of the right quality and quantity can be made. It will confirm that the technology is one which his organisation can handle and operate safely and competently, and will highlight areas where this might need specific attention. It does not tell him details of how the technology works (the chemistry, the physical steps, the special designs, the key items of equipment, catalysts, special processing techniques). It does not necessarily tell him what the efficiencies of individual unit stages in the process sequence are, but will give him across-the-plant efficiencies. There are many black boxes where the essence of the technology is hidden but he does not need to know what is inside them in order to make his technology selection.

If the technology has not been licensed before then the prospective licensee will need to be doubly sure at an early stage that the technology is translatable from one organisation to another who is 'an intelligent novice' (eg it does not depend on personal skills for utility) and the industrial and social infrastructures are comparable in the two different manufacturing locations and markets.

All this information is important to the decision-making process, but its utility may stop there in the sense that it will be of little or no practical use. Even if some information should be disclosed that is technically useful, its utility may be limited to the particular technology context of which it is part (ie the particular licensor's technology) and so possessing such information need not be an embarrassment to the prospective licensee should he choose another licensor's technology for his manufacturing venture.

CONFIDENTIALITY AGREEMENT TERMS
What does a secrecy agreement for assessment data usually say?

LICENSOR'S DISCLOSURE OBLIGATIONS
Sometimes the agreement merely recognises that the prospective licensor will be supplying information of a certain class, without legally obliging him to do so. More usually, the prospective licensor undertakes to supply information of a certain class but only that which, in his sole judgement, will be sufficient to

enable the recipient party to make a preliminary assessment of the technology and to determine his interest in acquiring a right to practise the technology. In some cases, the agreement will define by type, category, and depth of treatment, as well as form of presentation, the information to be supplied.

In some situations, there will first be a presentation (in the form, say, of answers to a questionnaire) of the prospective licensee's requirements and circumstances as they would dictate the basis for any design of plant. Additionally, there may be testing of raw material/feed stock samples and evaluation of typical products.

LICENSEE'S NON-DISCLOSURE OBLIGATIONS
These are always present even when the information to be supplied will have no practical utility but will be relevant only to a decision-making process. The non-disclosure obligations will consist of an undertaking not to disclose to other persons (individuals or companies) any received information and, perhaps additionally, the fact that the technology is being evaluated. There may be an obligation to confine received information to those regular employees, officers and directors who reasonably need to have it for the purpose of the evaluation.

Occasionally, individual recipients are required to countersign a copy of the agreement to acknowledge their understanding of their responsibilities. Sometimes, but not often, licensors insist on knowing who these individuals are. There is merit, in suitable cases, in limiting confidentiality to information supplied in written form, or promptly confirmed in writing.

The non-disclosure obligations should expressly not apply to, or should cease to apply to, information corresponding in substance to:
(1) Information already in the public domain by publication or otherwise (eg discernible by study or analysis or dismantling of things publicly available).
(2) Information subsequently coming into the public domain except by default on the part of the recipient, his servants or agents.
(3) Information which the recipient can show was in the recipient's possession at the time of receipt of the evaluation/assessment data, being information which is at the recipient's free disposal. (It can be most important for licensors to ensure that this information is still regarded as confidential unless and until the recipient makes a public disclosure of his similar information so that the right of the supplier to patent it, if he wishes, is not defeated by his disclosure to a recipient who already possesses similar information but, of course, does not admit to it.)

(4) Information lawfully acquired by the recipient from a third party and which the recipient is no longer required to keep secret under the terms of acquisition from the third party. (Sometimes it is stated that the third party shall not have himself acquired the information directly or indirectly from the licensor.)

One further exclusion should always be considered at the technology selection stage. It is information which has been developed within the recipient's organisation after receipt of the Licensor's information by persons who did not have knowledge of or access to the received information, or, alternatively, by persons who did not use or materially rely on received confidential information in the planning and execution of the development that generated that information. This provision is a shield not a sword to use against the Licensor. What it does do is seek to exempt duplication that has no causal connection with received information. If development of similar technology on similar lines is going on in the recipient company, but lagging behind perhaps in some or many aspects, the prospects for eventual honest duplication are real, certainly within the life of many confidentiality agreements. It ought to be possible for evaluating companies to avoid undue prejudice by some such provision backed up with tight administrative segregation of information and security practices. It involves 'proving a negative', but the evaluator should give himself a clear chance, in good faith, to answer effectively the charge that, but for receipt of the licensor's information, he would not have pursued this or that line of development or made this or that development. This added provision is particularly important when restrictions of use obligations are considered.

RESTRICTIONS ON USE OF RECEIVED INFORMATION

These ordinarily consist of a straightforward undertaking to use the information for the purposes of the evaluation and not otherwise. Again there must be exclusion of independently available information in categories 1 and 2 above and, if possible, independent developments should be excluded. Two further exclusions are needed.

First, a recipient should not be denied by contract the right to use as he pleases information already in his possession which was developed by him, however similar it may be to information he receives from the licensor. (An existing freedom to disclose that information and freedom from patents to use the information are not necessary additional qualifying conditions.)

Secondly, a recipient should not be denied the right to use information acquired by him at any time from a third party in whatever ways his arrangements with that third party allow. Sometimes, it is stated that the third party information should not have been obtained by the third party directly or indirectly from the licensor. This rider, which was also mentioned under non-disclosure obligations, will ensure that disclosures by contractors or other licensees in the course of discussions of previous experience of the licensor's technology will not defeat the letter of the Secrecy Agreement.

On occasion, the licensee's information which is given to the licensor has sufficient commercial sensitivity to warrant reciprocal non-disclosure obligations on the part of the licensor.

GENERAL PROVISIONS

These may deal with allowed disclosures to consultants/contractors and Government Agencies, and the terms governing such disclosures. The right, on terms, to pass the information to the licensee's parent or to subsidiaries may be conceded. A release to make disclosures required by a Court in legal proceedings may be given.

They may also deal with procedural matters such as the return of information after decision or after a set time period or even on demand, express limitations on copying, the right in any event to retain one copy of record in corporate confidential records, and reporting back the results of the evaluation.

They often stipulate ('for avoidance of doubt') that no right or licence under any patent or patent application is implied or granted by the evaluation agreement.

The parties should always consider placing a time limit (a back-stop date) on 'non-disclosure' and 'restriction of use' undertakings. Perpetual obligations are a legal and administrative nuisance. The shortest time that reasonably protects the licensor from prejudicial use or disclosure of his confidential information is the minimum period; say 5 years for economic assessment data. A reasonable period in most cases, bearing in mind that technologies keep advancing or get replaced, and recognising the limited practical utility of information supplied to the potential licensee at the assessment stage is 10 years, exceptionally 15.

Reference has been made to information already in the possession of the licensee and sometimes the contract requires that it should exist in documentary form or be evidenced in writing if it is to qualify for exemption. A prudent evaluating company would, so far as is reasonably practicable, ensure that its relevant existing information is already properly written up and accessible. A dated sealed deposit with an external party such as a Commissioner for Oaths or Notary Public can prove to be a most valuable precaution.

DESIGN PACKAGE AND LICENCE OPTIONS
On occasion, a potential licensee may need to purchase a front-end engineering package for the plant he has a mind to build, with suitable licence options available that he can take up on demand.

A capital intensive investment might justify acquisition of a full design package as part of pre-sanction evaluation and so might an investment confronting a constrained technical, operational, patent, environmental or economic situation.

The principles which apply to confidentiality and restrictions of use are the same as those discussed above, but now a complete, fully collated, package of useful information is passing that is specifically tailored to the licensee's needs and stiffer security measures and longer term obligations will be expected.

The design package will be the subject of a design contract. The one extra requirement is a clear option to a licence of appropriate scope and on acceptable terms. The full licence need not be negotiated in all detail but it could be, and certainly the main elements must be settled. The period allowed for exercising the option should be agreed. A significant payment for the option, and for any right to extend the option, is a reasonable requirement, if the existence of the option would be an actual deterrent to others seeking a licence or would exhaust the marketability of the technology in the evaluator's territory.

Detailed assessment of the patent position, which sooner or later is always required, can now readily be carried out since details of the technology are known. This will more particularly be concerned with the third party patent infringement aspect but will also have a bearing on the value to the licensee of the licensor's patent rights both as protection against unlicensed competition and as leverage for the fees being charged. The ultimate licence does not have to conform to the option terms; the option is the licensee's fall back entitlement.

Visits to other plants operating the technology will be a key part of this full evaluation. Training needs can now be sensibly discussed.

All these activities have as their aim the minimising of technical and commercial risks. But suppose that, after all, the evaluated technology is not adopted, the option is not exercised, and, sooner or later, the evaluator decides to go ahead with a similar investment based on his own or another licensor's technology. He will be exposed to an inescapable risk that the rejected licensor will charge him with misuse of technology in breach of the undertakings originally accepted. This charge might be motivated by honest concerns or be mischievous or malicious.

The immediate consequences are time-consuming and costly legal proceedings. If the charge is proved (and subconscious copying is possible amongst people of integrity), there will be damages and possibly an injunction. More serious still, in many ways, is the long term effect of getting a public reputation as a company which cannot be entrusted with confidential information and which ignores inconvenient obligations when it suits its purpose.

The true position may be that there has been no copying of earlier designs and no misuse of information received from the rejected licensor in breach of continuing contractual obligations. Nevertheless, it is difficult to stop the proceedings turning into a fishing expedition by the rejected licensor who is wishing to identify where the deficiencies are in his technology and what the improvements are in the adopted technology. The difficulty of persuading the successful licensor to agree to his technology being disclosed to the other licensor's nominee or even, in some legal systems, debated in open court should not be minimised. Such consent will be necessary for a proper defence. If technology developed lawfully in-house is being used, it will be necessary to disclose what that technology is and the route by which it was developed.

VISITS TO LICENSEES CAN BE VERY REVEALING

The disputants may, of course, agree to submit the issues to arbitration or expert determination in private proceedings, and, if so, to accept that the arbitrator's or expert's decision will be final at least as to whether and to what extent there has been a misuse of the rejected licensor's technical information. These are all very difficult issues and there are no easy solutions.

Sometimes an evaluator company will elect to distance itself totally from the detailed evaluation of technologies by employing a consultant in whom it has the highest confidence to make the assessment for it and, in the light of available in-house alternatives, make the technology choice for it. Recently retired employees have been used in this capacity.

4. THE COMMON TYPES OF LICENCE AGREEMENT

In Chapter 2 the place of technology licensing as an element in business strategy was considered. In Chapter 3 the factors inherent in the evaluation phase prior to offer and acceptance of a licence were discussed. In this chapter the common types of licence agreement are reviewed to highlight the aims, objects and effects of their usual principal elements. Chapter 5 considers the structure and wording of licence agreements. Appendix II discusses the licence grants in greater legal detail. Appendix III contains annotated sample agreements.

Licence Agreements fall into one or other of three general types. One type is licences under which rights pass in intellectual property (eg patents and registered designs) but there is no transfer of confidential technical information. Another type is licences under which confidential technical information passes for use but no specific licences for intellectual property are involved. The third type occupies the middle ground and combines and develops features of the other two, being licences where confidential technical information is supplied for use and, in addition, rights in intellectual property are granted expressly in the formal agreement or by necessary implication from the supply of goods, plant, or designs or from the conduct of the parties. Within these general types, there are sub-divisions which are conceptually useful. The discussion which follows is limited, so far as it involves intellectual property, to patent licences, and covers:
- Bare Patent Licences

(i) Specific Patents

(ii) Field of Activity
- Supported Patent Licences
- Implied Licences
- Information Licences
- Technology Licences (Patents and Information)

(i) Plant (Unit) Licences

(ii) Production Capacity Licences

BARE PATENT LICENCES

SPECIFIC PATENTS

This is the simplest type of licence. The patents and/or patent applications in question will be identified and listed by official serial number.

(a) EXCLUSIVE OR SOLE LICENCES

The licence will say whether and in what respects, within the ambit of the rights granted, the licence is to be exclusive (ie to exclude even the patent owner) or to be sole (ie no-one else to be similarly licensed by the patent owner) or to be non-exclusive. The European Commission, as the administrator of Articles 85 and 86 of the Rome Treaty, looks critically on exclusive and sole licences which have a Common Market dimension. Exclusive selling rights are more contentious than exclusive manufacturing rights and in any case, within the EEC, are heavily circumscribed by the 'exhaustion of rights' doctrine discussed below.

As a practical matter, 'exclusive' or 'sole' licence rights should be confined to those activities which need protection (to justify licensee's risk and investment) and should be of limited duration, eg 5 years from first commercial operation. In most cases, protection against competitive manufacture in the territory for which the licensee has manufacturing rights will suffice. If that territory is the whole of the Common Market, then exclusive or sole selling rights might be desired also. However, once goods are put on the market within the EEC, by the patent owner or through the exercise of a patent licence, their free circulation to other parts of the EEC cannot be prevented by other national patents of the patent owner covering those goods or the licensed process of manufacture. This is the doctrine of 'exhaustion of patent rights'. The doctrine does not extend to allow the use of the patented goods in other countries of the Common Market in special processes or applications that are themselves additionally patented by the patentee in those countries.

'THE PURPOSE OF AN EXCLUSIVE LICENCE IS TO PROTECT THE LICENSEE..'

Since the purpose of an exclusive or sole licence is to protect the licensee, the licence should desirably empower the licensee to grant sublicences in the field and territory of his privileged position. Failure to provide for sub-licensing is to court objection on competition grounds, since neither licensor nor licensee are contractually free to grant another party any licence. Unless a short period of exclusivity is built in to the licence (after which the rights are converted to non-exclusive) it is desirable to lay down conditions such that the Licensor will have the right to revoke the licence, or convert it to a non-exclusive licence, if there is insufficient working of the licence by the licensee. The definition of what is to constitute 'insufficiency', and allowed excuses, is always a challenging task especially where the licensee has to create the market for the relevant products. Conversely, licensees may seek the right to convert the licence to non-exclusive and thereby escape lump sum payments or minimum annual royalty payments which constitute the 'premium' for exclusivity.

(b) LICENCED ACTS AND TERRITORY

The licence may be for manufacture, use, import or sale of products or goods. It may be for any combination of the above. The licence may relate to patents in one country only, in certain countries only, or in all countries (for example for sales of products and goods).

A common example of a bare patent licence will be one for manufacture in a given country, or at a particular facility in it, use and sale of the products or goods in that country, together with exports. Frequently some export territories will be excluded by the licensor retaining his right to enforce his patents in those excluded territories against imports, assuming the relevant patents cover the imports or the relevant patent laws regard importation of substances made abroad by a process that is patented in the importing territory as patent infringement. But parts of the Common Market may not be excluded in this way.

It is a self-evident consequence of a bare patent licence that patents or rights not enumerated are not licensed. Thus, a licence merely to make and use a patented product is not a licence to sell it, and a licence merely to sell a patented product is not a licence to make it nor a licence under patents covering special downstream uses of the products. Such licences do not imply licences under patents covering product components or ingredients or methods for their production.

(c) LICENCE SUBJECT MATTER

The patent licence may be for everything and anything within the scope of, and covered by, the patent claims (which define the subject matter of the patent monopoly). Alternatively, it may be for specific embodiments only, that is, certain members of the class of things, processes, or methods collectively covered by the patent claims. There is a problem with attempts to confine the exercise of the licence to certain uses only of the patented goods or products except, of course, where the exclusions are uses that are themselves patented or the industry is already structured into distinct domains. To attempt to do this may be challenged as a patent mis-use (at least in the Common Market and the USA). It can be seen as an improper attempt to 'divide and rule' areas of activity outside the actual scope of the patent rights. The fact that without a licence to make and sell the patented items neither the licensee nor anyone else could use them for any purpose (since they could not lawfully get hold of them) is not seen as a justification for upholding restrictive 'applications' licences.

It is to be hoped that objections to field of use limitations do not lead to them being regarded as *per se* violations, because circumstances can be envisaged where they are reasonable, in the public interest, and do not restrict actual or likely competition. An example would be a new chemical or manufactured substance such as a biologically active compound, new alloy, plastic or ceramic that has a number of potential uses in diverse industries and where the necessary development effort and expense to bring it to commercialisation justifies exclusive or sole rights for each of those industries.

(d) SALES NON-ASSERTION

A patent licence may include undertakings by the patentee that he will not assert any patents he may have in given export territories to prevent the licensee's customers either importing patented products or downstream derived products into these countries or using such imported products in particular patented applications. Such undertakings rely for validity and enforcement on contract law (and the doctrine or privity of contract, which is a feature of English law, can deny the customer the legal right to enforce such an undertaking although given for his benefit). Such undertakings are vulnerable to a change of ownership of the relevant patents in the export countries or their exclusive licensing for those export territories. Nevertheless, such undertakings are a proper and convenient aspect of patent licensing and, once acted upon and known in the trade, it becomes difficult for the intended beneficiaries to be denied the benefit of them.

It must be noted that this is not the same legal situation as where the licensee has rights to sell patented products into specified export territories and additionally, say, rights for use in those territories in patented applications coupled with a right to grant sub-licences to his customers for onward sales or use. The sub-licensed customer acquires a legal right.

(e) DURATION

The patent licence may be for the entire remaining lives of the relevant patents or for a specified lesser period if that is mutually acceptable. Usually the licence will be for the full term of the patents then in existence (including pending applications) unless terminated by the licensee at an earlier time by formal notice in the terms of the licence agreement or by the licensor for an unremedied material breach of licensee's obligations.

(f) PROBLEM ISSUES

One particular area in patent portfolio licensing (ie licensing a bundle of patents) that is a problem in the USA and is being questioned in the EEC is fixed term flat rate licensing that takes no account of the differing expiry dates of relevant patents and does not distinguish major from minor patents. Another questioned practice is levying the same royalty whether a licence is taken for a whole portfolio or for certain selected patents only. A further practice that is frowned on is discriminating between different licensees by levying substantially different royalties if they are in fact competitors, unless of course there is an objective justification such as a discount to a first licensee who has borne the risks of market development. Another point to be careful about is requiring royalties to be paid on products of a kind or class that is not defined closely in terms that bring it wholly within the scopes of the licensed patents. Administratively this can be convenient and quite satisfactory to the parties but the divergence from the scopes of the licensed patents should be minimised if it is not to be vulnerable to challenges. It is not acceptable in the European Community to require a licensee by contract to waive his right to challenge the validity of the licensed patents but some national legal systems may in fact place such a restraint on a licensee by the doctrine of estoppel. Within the EEC it is considered permissible for a licensor to insist on a contractual right to terminate the licence if the licensee should challenge the validity of the licensed patents as by an action for revocation. It is acceptable to levy and retain royalties on all products covered by the claims of the licensed patents even though those claims might perhaps be struck down if revocation proceedings were taken. A licensor should not, however, levy royalties on patents which he knows to be plainly invalid. Both licensor and licensee may see real potential for further development of the patented technology or innovative applications of it. The treatment of future patents for yet-to-be-devised improvements and developments (and the effect on royalties) needs great care to avoid unacceptable impositions or restrictions, while capturing the evident benefits to both parties.

Later in this chapter, the financial terms of licences are discussed in some detail under a specific chapter heading.

(g) SPECIAL POINTS

A patent licence should always address the question "what if the licensed patents terminate, lapse or are revoked?" In some countries, the patent law will provide part of the necessary answer but a clear contractual statement is nevertheless desirable. The licensee should also write in provisions which cater adequately with proper maintenance of patents (if that is important to him) and which prescribe actions, suspension of royalties (or some portion) and rights or obligations to pursue infringers if he should encounter substantial infringing competition. Under some patent laws, exclusive licensees have the right to sue infringers direct but a wider class of licensee can obtain by contract the right to pursue action against infringers (eg by suing in the name of the patentee). This is however not a recommended policy in the USA.

If the licence refers only to pending patent applications, it is necessary to deal satisfactorily with the situation if the pending applications are withdrawn before publication so that the revealed inventions stay secret or if the patent applications are abandoned before grant.

(h) MINIMUM OR MAXIMUM SCALE OF USE

A licensee, especially one with sole or exclusive rights, might be required as a condition of continuing privileged licensee status to make a minimum number or quantity of goods in any agreement year or alternatively to pay royalties as if he had. Unless there are very special factors, no licensee ought to accept, and no licensor should seek to impose, a limit on the maximum use which the licensee may make of his licence during any year or in aggregate. Of course, if the licence relates to a particular plant or a small isolated market, there may be other reasons why there will be a practical or economic limit on the use the licensee can make of his licence in any year.

FIELD OF ACTIVITY LICENCES

A patent field licence is a contractual document which says that the licensee is granted the right for a given territory to do specified acts in a specified business area (the 'field') despite those patent rights which the licensor owns and controls and which relate to or cover that field and which would or could be infringed by doing the specified acts in the territory. The field might be a type of product or a class of production or manufacturing process. The specified acts could be manufacture, sale, use, export, etc. The point of a field licence is that it is the contractual terms which determine the 'rights' and not solely or predominantly the patent scopes. If the licensor has one, several, or many patents that cover the 'allowed' acts, the licensee is unconcerned — he has his permit. It is essential to define the field with some precision; it is also of the utmost importance to set out very closely what acts and enabling steps are allowed. Licensors will be very concerned to ensure that a field licence does not act as an open door through which rights in distinct proprietary patented technologies flow to the licensee for his use in the agreed field of business activity.

The business activities that give rise to payment obligations, eg manufacture plus either use or sale, will be specified contractually and the unit of operation or the class of product that qualifies for royalty payment will be defined as will the royalty rate per unit of operation or per unit of weight of product. This is the case for all patent licences unless of course the payment is solely a lump sum fee (in lieu of royalties) for both the privilege of permitted use and actual unlimited use. Obviously these are matters for careful definition. In the case of a field licence, a licence fee or royalty is plainly in the nature of a levy on an area of business activity and the question whether all goods or unit operations that contractually qualify for royalty payments are in fact ones that would, but for the licence, infringe some valid patent or patents of the licensor is a major factor in the drafting of payment obligations to the satisfaction of both parties. In the case of licences under enumerated patents only, it is usual to find that the acts or products qualifying for royalty payment are closely defined in terms that echo the scopes of the claims of the specified patents for the reasons suggested above. This is less likely to be the case with a field licence under a portfolio of patents.

SUPPORTED PATENT LICENCES

The essence of these is still the patent licence, and all the above comments apply. However, a patent holder will often have research reports on the fundamental science, reports of laboratory experiments and perhaps pilot studies, as well as market development information. All these, or some of them, he may be willing to supply to a licensee to assist the licensee to achieve earliest beneficial use to the advantage of both parties. Whether or not a charge is made for this assistance, and whether or not the information supplied is to be kept confidential are matters which will be decided on the facts. If there is to be confidentiality and contractual limitations on how the information may be used, then the transaction begins to look more like a full technology licence and the issues discussed below under INFORMATION LICENCES and TECHNOLOGY LICENCES will need to be faced.

IMPLIED LICENCES
Two kinds need to be mentioned:
- user licences that arise by contractual implication from the terms of contracts for the supply of goods ('in aid of sales');
- user licences that arise by contractual implication from the supply of plant to be used for given purposes, or from the supply of designs of such plant ('vital to the purpose of the supply').

IN AID OF SALES
Many suppliers of products provide to their customers technical service information aimed at facilitating best use of their products. Plainly the assistance offered and accepted is intended to result in the information being used; there is an implied licence, at least for the intended purpose. The supplier may have patents covering downstream uses. Since royalties are seldom as rewarding as margin on added sales, the patentee supplier may choose to offer his product to customers at a competitive market price on the understanding that customers will be using them in the patented application. He impliedly waives his right to enforce his patent in respect of such use of those specific products. This is a sort of label licence. (An explicit label licence arises when the container or the article itself or the document of sale carries a statement that buyers and users of the product or article in question are licensed to use it under enumerated patents. A notional royalty is included in the sales price.)

People have always been a bit uncomfortable about label licences without being able to articulate the legal basis for their worry. Under many patent laws, anti-trust and competition laws, it is unacceptable for a licensor to require (ie oblige by contract or other pressures) a licensee under a patent to obtain his requirements of non-patented goods from the licensor in preference to other available sources.

But this is not quite that situation. Nor in most cases could it be said there is abuse of a dominant position. In any case, there is no legal prohibition on a licensee freely electing to obtain his requirements on bona fide negotiated terms from his licensor. The law does not force him to buy elsewhere. Indeed, there are many fields in which the choice of seemingly equivalent components, ingredients or raw materials can involve unquantifiable risks such as in electronics, effect formulations, fibres and pharmaceuticals.

Perhaps it is not the label licence as such that is the problem, but rather any subsequent actual refusal without objective justification to offer a patent licence on reasonable terms to a customer so that he may, if he wishes, use competitors' products.

PLANT AND DESIGNS SUPPLY — 'VITAL TO THE PURPOSE OF THE SUPPLY'
Suppliers of plant designed for specific jobs and contractors who supply designs for such plant commonly do not expressly license the buyer under patents they may hold covering the operation of the plant. An implied licence is acquired by the buyer/operator, but in respect of that plant and the intended operations only. Commonly buyer/operators are entirely happy with this situation but if key plant components, consumables or catalysts are themselves patented or the plant can be made to operate in a range of patented modes or can be modified readily for use in other patented ways then clarification of the situation, as can only be achieved by express language, is needed. Appendix II develops this theme further.

INFORMATION LICENCES

THE BENEFIT
Technical information that is useful to another for the better understanding of, or development of, some existing product or process is a tradeable item. The benefit to the acquirer is either the economic impact of the information on his business less the price paid (ie its utility value) or it is the saving in cost, time and effort which would have otherwise had to be expended in order to generate equivalent information independently (ie investment saving).

The exchange of research-, development- and operating information and experience is a normal business activity in many industries, even between companies who compete at the same market level or internationally in the same markets for the product to which the exchange relates.

A balanced exchange will lead to greater efficiency and better general competitiveness without, it is presumed, adversely distorting the relative competitiveness of the exchanging parties. Anti-trust laws, EEC competition laws,

'TECHNICAL INFORMATION THAT IS USEFUL TO ANOTHER FOR THE BETTER UNDERSTANDING OF, OR DEVELOPMENT OF, SOME EXISTING PRODUCT OR PROCESS IS A TRADEABLE ITEM..'

and national competition laws all encroach on these arrangements, especially if one of the parties is a major force in the market for the supply of the relevant product.

PATENTS

Even though the parties may each hold some patents in the relevant technology area it is, in mature industries at least where no dominating or 'master' patents remain, not unusual to see technical exchanges or one-way technical disclosures agreed to without rights being granted under patents. Indeed, the agreement will often say that no right or licence under any patent rights of either party is granted or implied by the agreement, lest it should be presumed, or argued later, that the minimum patent licences necessary for fullest immediate use and product sales were granted by implication. Internationally it is still legitimate to protect market territories against exports by licensees by withholding, and if necessary asserting, rights under patents in those territories. Even within the Common Market where, by rule of law (the 'exhaustion of rights' doctrine) free movement of goods cannot be prevented by a corresponding patent in a market country, this is not the same as a licence to the licensee to do it.

If some form of licence under patents is to be a feature of the agreement it may be expressed, not as a patent licence proper, but rather as an undertaking not to assert patents against specified acts using specified information in specified territories. The patent clauses of information licences or information exchanges (which are cross-licences) are often expressed in this format and great care is needed to ensure that they are clear and workable. The crux is the proper definition of the information whose use enjoys freedom from patent suit. This topic is developed later in this chapter and in Appendix II, but the following example may illustrate a few of the problems.

Two manufacturers, A and B, use similar production technologies. They decide to exchange details of their technologies. In practice, all exchanges are incomplete but, more importantly, the two teams who are effecting the exchange will gloss over areas of similar technology and focus their attention on respective novel features. Information that is in fact formally exchanged will be less than what the agreement allowed for. If the patent clauses were expressed merely as a freedom to use received information, then A will acquire freedom to use B's novelties disclosed to him and vice versa, but what, say, will be A's position in respect of any patents B has that obstruct A's exploitation of A's novel features or even that cover subject matter in the unexchanged common ground? Thus, by the manner in which the teams choose to implement the information exchange (and very sensibly too) the parties do not achieve the patent freedoms they no doubt intended and, indeed, they achieve the curious result that A may be free to use B's improvements, and B free to use A's, but neither may necessarily be any freer to use its own. This problem is not solved by linking the patent freedoms to information to which either party is entitled irrespective of whether it formally passes. A cannot know what B might have told him but did not. The safe solution is to exchange free-standing patent field licences (couched in the language of the information exchange and the agreed fields of use).

CONSIDERATION

Information licences tend to rely on lump-sum payments, or cost-related payments, or consideration in kind (ie exchange of information) or even if tax regimes will cooperate on goodwill generation as the reward to a party for his supply of information and the effort involved in collating and handing it over. Royalties or fees related to 'extent of use' or 'intrinsic utility value' are inappropriate when some only or none of the information supplied may be of actual immediate practical use to the recipient. The utility may rather be in the nature of confirming what the recipient already suspected or had experienced or of highlighting what to avoid doing. It may merely anticipate what the recipient would have learned from R&D programmes already begun or planned. Additionally, the passing of commercial information on scale of operations or product pricing that royalty schemes entail will usually not be acceptable to parties engaged in this kind of licence.

KEY FEATURES
Information licences are distinguished from full technology licences because, unlike the latter, it is their purpose to supplement existing production knowledge and not to provide complete information for a new production venture. However, the licence terms will be conceptually similar to the 'information' terms of a full technology licence discussed later.

In summary, the key features of information licences are:
- technical definition of the field of information supply;
- the time frame both in respect of information that qualifies for the supply (eg information developed or acquired between specified dates) and in respect of the time period over which qualifying information is to be accessible to the licensee;
- how the supply will be effected, eg initial package supplied, visits, meetings, 'top-up' on specific request;
- non-disclosure obligations and duration (as well as, of course, necessary permitted disclosures to consultants, contractors, and customers);
- limitations (if any) on use of supplied information outside the plant or business for which it is supplied, and their duration;
- whether any patent licences or immunities are given in respect of use or product sales, and if so what the scope of these is;
- payments, and their basis of calculation.

It is unusual for information licences to include:
- indemnities in respect of third party patents;
- guarantees or warranties in respect of plant performance or product quality.

The receiver ordinarily bears all the technical and business risks of applying received information.

It is often the case that the utility of information intended to supplement the licensee's knowledge is in fact self-limiting; it can be of practical use only in the context for which it is supplied. Contractual restrictions on use may therefore not be required by the supplier; he will be content merely to see adequate non-disclosure undertakings. A licensee will prefer not to have contractual user restrictions; he will want the licensor simply to rely on such patents as he may possess. There is much to be said for limiting confidentiality obligations to information that is either supplied in writing or if passed orally promptly confirmed in writing. The implications of intimate intermingling of two parties' information need to be faced up to and proper documentation is important both for confirming what was received under the licence and also for establishing what the recipient already knew or would inevitably have found out from current R&D programmes. It is important that the receiver of supplementary information should be practically as well as legally free to deal with the information he already had as if this licence had never existed.

TECHNOLOGY LICENCES (PATENTS AND INFORMATION)

OBJECT AND AIMS

Under this heading, licences are considered that have as their object the establishment in a licensee of a manufacturing plant or production facility whose design and operation are based on technology possessed and supplied by the licensor. As a minimum the technical means to realise the plant or facility and to operate it passes. That means may consist of information, eg a design package suitable for engineering workup and outline operating instructions, or it may include services such as operator training, start-up supervision or even hands-on operation with the licensee's staff shadowing their licensor counterparts during the initial period of operation of the licensee's plant. Sometimes it will include supply of key plant items, equipment, instrumentation, catalysts, control systems and computer software.

The aim of the supply is the secure, mutually convenient and timely equipping of the licensee to do something new and as to which he is presumed to have no special knowledge or experience.

It is rarely the object of such a technology licence to place the licensee on the same level as the licensor in terms of knowledge of the technology. Licensors are understandably reluctant to disclose to licensees the fundamental science and design data and correlations on which the licensors draw to produce specific technology packages supplied to licensees. Licensees are therefore rarely competent to do much other than engineer, build and operate the particular plant or facility for which the design package was provided, or duplicates of it (disregarding for the moment whether the right to do so has been granted). This is not to say that licensees never have technical knowledge and experience relevant to facets of the process technology or unit operations. Of course they may, and, if they do, they will wish to incorporate their own improvements. The technical risk of altering licensor's designs will be carried by the licensee and may void any guarantees, warranties or indemnities offered by the licensor or made them inapplicable, which amounts to the same thing. Responsible licensors will try to meet their licensees half-way by objectively evaluating the effect, if any, of the proposed modifications, and redrawing the guarantees, etc to fit the new circumstances. Sometimes, though, when licensors are dealing with unsophisticated and inexperienced licensees they understandably stipulate that their designs are to be used in total or not at all.

TYPES OF TECHNOLOGY LICENCE

Two main types of technology licence can be identified. The first is a licence tied to a particular plant to be built at a particular location; the second is a licence for production capacity in a given territory that finds initial expression in a specific investment but allows the licensee to expand the plant and add new capacity.

(a) PLANT LICENCE

A licence for a particular plant is a very limited and limiting transaction. The technology transfer will be just that needed to enable the licensee to build and operate the plant. Use of supplied information will be confined to the plant and its product(s); so will rights under patents. The rights will die with the plant, by and large. The right to enlarge the plant will be severely curtailed. Perhaps consent will be needed for any enlargement or only for enlargement of defined capacity-determining components of the plant. Increased throughput by driving the plant harder or by introducing operating improvements should not be prevented by the licence terms (even though there may be royalty payments to be made for any resulting increased output).

Particular points which licensor and licensee, from their different perspectives, will need to take into account are:

(1) Production plant design is not an exact engineering science. Design capacity is a target but gross overdesign of the proposed plant by the licensor, or his design contractor, will im-

pose a capital and operating cost penalty for which, guarantees apart, they are not liable. It is proper to consider whether achieved production beyond the sought-for nameplate capacity should necessarily give rise to the added burden of further royalties, and in any case whether a ceiling should apply at, say, 115% of nameplate capacity.

(2) Is there to be feedback of licensee's improvements and developments, and, if so, should it be for all information in the general technical field or only for direct operating plant experience? Should the feedback for a technical field include research information or only that which has been developed to the point of commercial utility? What grant-backs under future patent rights are sensible? For how long should this feed back continue? What confidentiality terms should apply?

(3) Should the licensor reciprocate with his corresponding information? Should the licensor be free to pass the licensee's feedback to his other licensees? Normally licensors insist on this right, but is it fair to require that other licensees benefit only if they too reciprocate on the same broad basis? No exclusive grant-back of licensee's patents and no grant to the licensor of sole licensing rights for licensee's improvements should be demanded or agreed, even though the improvements may have no utility outside the licensor's technology and even though disclosure of them to others by the licensee would require the licensor's consent since they are inextricably intermingled with licensor's information to which non-disclosure obligations apply.

(4) What freedom is there for direct licensee-licensee dialogue, or trade-offs, albeit with licensor present. What mechanism will ensure that licensees will have access to the pool of accumulating experience? A right to have a full and open dialogue between licensees within a group of connected companies and with their common major or sole parent should be fought for most vigorously. Anything less is an economic, administrative and legal morass.

(5) It is necessary that strict non-disclosure obligations should be relaxed to allow the licensee to have discussions with contractors and consultants, and to allow orders to be placed on vendors and suppliers. The terms which will govern such third-party dealings will need to be specified. Licensors should rarely seek to dictate to licensees whom they may deal with or to unduly restrict their choice, but licensors may insist that particular third parties who will need to have a substantial amount of information should be acceptable to them and should be bound by direct secrecy undertakings. In industries where particular technologies are widely licensed, licensors will provide lists of approved

vendors/suppliers (or their design contractors will) whose items or products have proven suitability, and with whom they have in place appropriate contractual arrangements concerning proprietary design data, inspections, tests, and cross-undertakings on performance limitations. This can be most beneficial to licensees.

(6) Consideration should be given to the possible conflict between non-disclosure obligations and a licensee's wish to be free to patent his improvement inventions.

(7) Should royalties be payable by the licensee on extra production achieved by applying licensee's improvements if they are ones that are within the obligatory feed-back of information to the licensor?

(8) At what point in time should confidentiality undertakings and restrictions on making other or further use of licensor's information sensibly expire? And, when they do, does the licensee have a clear right to make commercial use of the acquired information freely and for all purposes, subject only to licensor's patent rights enforceable against such other or further use? Licensees must strive for this clarity but should not be unmindful of a licensor's genuine concern not to put his licensee in the position of competitive licensor for 'his' technology at a future time when it could still be very competitive. One intermediate course that has been adopted is to lift confidentiality and use restrictions at a chosen future time, but to prevent the licensee then electing to keep the technology confidential and to license it himself for reward. An ingenious concept, not without its problems!

(9) The agreement should not be so worded that any subsequently agreed extension to information exchange automatically operates to prolong confidentiality and non-use undertakings for all information and automatically to prolong the period during which royalties are payable.

(10) Even a one-plant licence should allow for erection of a replacement plant if the initial plant should be destroyed. Indeed, although this is a one plant licence, the parties should consider the inclusion of options for further licences for additional neighbourhood plant in addition to the more likely requirement of options, or rights, to enlarge the agreement plant.

Rights to enlarge the agreement plant, and options for further plant, are worthless to a licensee if he lacks an effective capability to design the new facilities, and there is no entitlement to call on the licensor to produce and supply the required designs.

(11) If the licensor, or his nominee, is to supply key plant items, processing aids, raw materials or catalysts, the licensee must consider the security of future

supply, including effective back-stop rights to allow self-help if the licensor or his nominee is unable or unwilling to meet future requirements. However, a licensee dependant on his licensor in this way is inevitably vulnerable, since economic alternatives simply may not exist.

(12) Does the royalty structure adequately take account of the relevance and expiry dates (or lapse) of licensor's patents? There is merit in differentiating royalties for use of secret information from royalties under patents since their payment is predicated on quite distinct legal bases. Additionally, they may attract different tax treatments.

Licensees should realise that if a production unit is designed by the licensor (so that it will incorporate specific collated confidential information fixed in hardware or operational modes) royalties related to use of licensor's information will be due for the operating life of the unit unless expressly stated otherwise. The normal 'gateways' of confidentiality clauses never reach that deep.

(13) What are to be the legal responsibilities and liabilities of the licensor or his contractor if the plant does not work, or works only very badly? What is to be the position if third party patents intrude? As prevention is always better than cure, the licensee should assess the third party patent position himself and not merely rely on the representations of the licensor. To some extent, of course, risk cannot be entirely eliminated. Visits to other users of the technology before major funds are expended can be very revealing. A licensee must be especially careful if the licensor is not the originator/owner or operator of the technology but is his licensing conduit such as a contractor. The terms of remit allowed the contractor, his breadth and depth of knowledge of the technology or the patent position, and his competence to produce designs to meet the licensee's needs and to provide effective commissioning or operational support and trouble-shooting may all be very limited.

(14) If rights under licensor's patents are expressed merely as an undertaking by the licensor not to assert his patents against any use of the licensor's information in the given plant or against any of its products, then as discussed earlier problems can arise.

This is rarely an acceptable position. A licensee may wish to substitute, modify, or improve upon the technical information of the licensor and he must ensure that freedom from suit continues. The only sensible course is to avoid the problem. This can be achieved by express patent licences, patent licences for defined technology fields, or blanket hold-harmless undertakings related to technology steps, processes or fields (irrespective of whose information is used).

(b) PRODUCTION CAPACITY LICENCES

These are important when the economic scale of production units is small relative to the expected market growth for the product or to the intended rate of growth of the licensee's market share for the product or downstream materials made from it. In other words, the licensee is in a business on a long term basis and intends to grow in it or to hold his relative market position in an expanding market. This sort of licence is also important for products that are new to a market or which have so far achieved low penetration. Here the licensee will prefer to build a small plant first and then add further capacity as business develops over a period of time that is substantially less than the economic or functional life of the first plant.

These licences have much in common with other licences discussed earlier. It is the special features that need to be highlighted.

(1) The licence contemplates as a natural development investment in additional capacity and this will include new plants. New plants will likely be different in scale from the first plant and will need to incorporate developments in the technology resulting from experience and learning. It will be necessary also to accommodate changing product specifications as the market becomes more structured and as competition stiffens. This means that the question of the licensee's competence to design or to obtain access to designs for new plant and plant enlargements and modifications is critical. A right to do these things without the ability to do so effectively is an empty right.

(2) Clearly, the question of where new plants will be situated is vital. The licence will ordinarily be for manufacture in a given territory or territories. The proper definition of that geographical area is vital to a licensee so that markets can be served by plants logistically well situated. If foreign manufacture is likely to be desired, it is necessary also to consider who would actually own and operate the plant. A foreign subsidiary or affiliate of the licensee may need to be used and, if so, the right to sub-licence (or the right to have that overseas operation directly licenced on

'WHERE PLANTS ARE SITUATED IS VITAL...'

pre-set terms) is vital. A novel modern species of joint venture is a jointly owned asset: a production plant owned in undivided shares by two or more companies. Plainly, it is not enough merely to arrange a licence agreement with the company who will operate the plant. Equally, all owners do not need an operating licence or need to have total knowledge of and access to the technology. But all will need some information, all will need to be bound with confidentiality obligations, all will need to agree to any improvements exchange, and all will have to guarantee royalty payments. Inevitably these situations are unique and complex, but the well-tried principles of licensing agreements are equal to the challenge.

When sub-licensing is permitted it is likely that the 'head-licence' will make the head licensee liable for defaults by sub-licensees (eg breach of secrecy undertakings or mis-use of confidential information) as well as requiring the head licensee to account directly for, or guarantee, royalties due on sub-licensees' operations. Tax questions become important.

More disastrous potentially is the consequence to licensee or sub-licensee of breach of the licence/sub-licence terms by either of them. It is usual for licences to give the licensor the right to cancel the head-licence if there is a significant and unremedied breach, whether or not the proper law would entitle the licensor to terminate the agreement for such breaches in the absence of such a contract provision. It is important, of course, to a licensee to ensure that termination for breach cannot lightly be imposed but it is equally important to ensure a life-line for innocent parties. An entitlement to substitute direct licences from the licensor to a sub-licensee on the same terms should be sought if the head-licence falls. The ability to continue the head licensee's own operations under the head-licence if a sub-licensee defaults (irrespective of any damages payable or any forfeiture of the right in future to sub-license others) is an important need.

(3) Royalty arrangements, as between licensor and licensee, need most careful definition (apart from questions of withholding taxes and double-taxation treaties). A fuller discussion on royalties and licence fees appears later in this chapter but immediate issues include the following.

The opportunity to aggregate production of all plants (including sub-licensee's plants) and to pay royalties for added increments of aggregate capacity on a descending scale should be considered.

The period of time over which royalties are payable needs specifying carefully. If royalties cease on given plants after so many operating years but they continue in operation, will their nameplate capacity, their actual output, or their highest or average life annual capacity be creditable against future royalty payments under an aggregation scheme? If licence fees for 'paid-up' plant capacity are payable instead of running royalties, are the fees payable on sanctioning the new capacity or on commissioning the new capacity? When do confidentiality provisions and patents expire in relation to fee obligations? A duty to pay full-scale fees for a new investment close to the expiry dates of patents or secrecy obligations can on the day seem needlessly expensive. When 'paid-up' plant ceases production and new plant is brought on line, how is its capacity to be credited?

(4) A licensee must ensure that if another licensor's technology should be selected for a future plant (and provided there is no importation into the new plant of the first licensor's specific proprietary technical information and no working under his patents) no royalties or fees are payable under the contractual payment terms of his first licence. Reject all 'springboard' arguments. Licensing always puts a licensee on the learning curve; that is the inevitable 'price' to licensors. If they fear it, they should not license.

(5) Exchange arrangements for future developments and improvements in the technology assume particular importance. Consider the licensee's 'vicious circle'.

Suppose the licence does not provide for improvements exchange. There is then the risk that the licensee's technical position will fall behind that of other licensees' and even farther behind the licensor's. When new investment is needed, will economics dictate that 'top-up' technology be sought, and what will the terms be? If the licensee needs the licensor to design his new plant will the licensor be enthusiastic about supplying an out-of-date package (the licensee's only entitlement, perhaps)? If the licensee's technical position stays up with the others, how are his improvements to be incorporated in the new plant without making a probable gift to the licensor who executes the design?

Suppose now that the licence provides for long term improvements exchange and obliges the licensor to provide design services for new plant. What effect will this have on the period over which royalties or licence fees are payable? Will the right to use what may be minor future inputs from the licensor cost the licensee full scale substantial royalties for later investments that might otherwise have been royalty free (or at a much reduced rate) because confidentiality terms and the royalty period in respect of the initial input technology would have expired or been close to expiry, and also free grant-back of perhaps significant licensee advances expensively won?

(6) The proper scope of patent licences and rights (especially any hold-harmless undertakings) as well as restrictions on the licensee's field of use of licensor's technical information are obviously vital issues since they must anticipate future production needs as well as process- or product technical developments and not merely the technical basis of the first plant or the type of product to be first made.

(7) A production licensee may consider that a period of exclusive or sole exploitation rights is important and the licensor may accede to this. If so, the patent licence should be so worded and additionally the licensor may undertake not to use himself (if appropriate) or supply to others for their use in conflicting ways the licensor's technical information relating to the process or products to which the licence relates. Legally, it is best if exclusive and sole rights are of short duration, that avenues for sublicensing are considered and that reversion of the licence to non-exclusive status can be imposed if the licensee does not adequately exploit the licence.

PAYMENT STRUCTURES OF LICENCES

The payments structure in a licence agreement does not influence the nature of rights in information or intellectual property which are offered or sought, but the levels of payments proposed for alternative parcels of rights will, of course, differ. Similarly, the payments structure may determine when agreement provisions are triggered.

What the payments structure does do is affect cashflow, financial risk, and the administration of the licence agreement. Licences can be royalty bearing or fixed fee or a hybrid which overlays royalty payments on a base of fixed lump sum payments.

ROYALTY BEARING LICENCES

Royalties are expressed in terms of either a fixed or calculable monetary sum multiplied by a number that is a measure of the extent of use of the technology, or a fixed percentage of a defined sales value (or, rarely, a standardised cost of production) for products sold or produced.

Royalties may become due when product is made, or it is sold, or it is used, as may be agreed.

Royalties due will be payable only at the times specified for payment; say within 60 day grace periods following end of calendar quarters, half years or years. Reports of relevant production, or sales or usage, for periods to which royalty payments relate will be called for at specified times.

Payments will be made by specified routings. Royalties may have to be paid net of tax at rates stipulated in the relevant tax regulations or they may, where regulations allow and necessary approvals are obtained, be paid gross.

Additional points are:

(1) The licensee pays only with use and in proportion to use. For this reason, licensor's services are made separately reimbursable.

(2) Minimum and/or maximum annual royalties are quite common, and sometimes a ceiling on cumulative royalties is agreed. Premium payments for exclusive rights were discussed earlier.

(3) Tax implications, especially when money crosses international boundaries, need careful study. A tax expert must advise and there is no substitute for up-to-date advice from an expert in the payor's country. It would seem that tax laws, practices, and even interpretations are ever in a state of flux.

However, it is worth stating that it is possible in many countries to place on the licensee the burden of local taxes and levies payable on royalties being remitted to the extent needed to ensure that the net (after corporation tax) retention by the licensor is what he would have achieved if he and the licensee had been in the same country. If this cannot be done, then royalties will have to be adjusted upwards or the reduced net revenue accepted. Payments for services may also be liable to withholding taxes and these can, in the case of services performed in the licensee's country, be very high indeed especially if the licensor's experts are on long-term assignments.

Licensors must establish at an early stage, what sorts of information supply and services attract which levels of withholding taxes, whether the taxes are regarded as taxes on licensor's income and merely paid on his behalf by the licensee or whether they are levies on the licensee's payment, what the quoted tax rates are a percentage of, what the double tax treaty position is, what credit the home tax regime will allow, and what paperwork is required. Not only do services rendered in the licensee's country run the risk of very high tax levels for which full credit may not be available but also the licensor risks being adjudged to be carrying on a business in the licensee's country — to have 'a permanent establishment' — and the tax implications are then disturbing.

(4) By and large, licensors insist on payment in hard currency. Internationally, US dollars, and to a lesser extent DM's, are favoured. But pounds sterling are still widely used.

(5) The period of time over which royalties are payable must be specified. In the case of a patent licence the period will be subject to continuance of the relevant patents and ordinarily the licensee will have the right to terminate the licence on suitable notice. Termination of any licence by effluxion of time will always bring payment obligations to an end (but royalties accrued up to the date of termination will be payable in the ordinary way). In developing countries, the period of royalty payments and their levels are matters of official interest and will ordinarily require approvals before the licence agreement can come into force; additionally, licensors may have to accept administrative delays at the Central Bank in the transfer of payments in hard currency.

(6) Inflation formulae designed to preserve the purchasing power of royalties received are fairly common except where the royalty is expressed as a percentage of sales value (which it is usually assumed will keep pace with inflation).

(7) Expressing royalties as a percentage of net sales value is quite satisfactory for arms' length sales by the licensee but special arrangements will be needed in order to deal fairly with products consumed internally and also with products sold to connected companies at transfer prices that may not reflect the external market value. Royalties expressed as a percentage of production costs are sometimes proposed but if these involve revealing actual costs, as opposed to assumed standard usages and consumptions and posted prices, then licensees will not often find this approach acceptable.

(8) A right to send in an auditor during normal business hours and, say, once a year in order to confirm correct reporting and accounting is required by licensors. The auditor should be authorised to report discrepancies only. The licensee bears his costs of the auditor's presence and the licensor pays the auditor's fees and expenses. If a significant discrepancy is revealed (say more than 5%, under-reporting) it is reasonable to require the licensee to pay for that audit and/or successive audits.

FIXED FEE LICENCES

This alternative approach involves a licensee in acquiring 'paid-up' annual production capacity.

Suppose a plant has a design, or 'nameplate', capacity of 1000 units (or 1000 tes) of product per year. This signifies that in an average year at expected levels of productivity and on-line time and with normal feedstocks and normal manning, the plant will turn out around 1000 units (or tes) of product. Of course, production plants do not have a unique capacity; what they can actually make in a year depends on a host of things, not least amongst which are the functional limits of duty of the component items of equipment, their reliability, and the inevitable design margins involved in all engineering design work. Design capacity is merely a target capacity under an assumed set of conditions. Plants are always either underdesigned or, more usually, overdesigned or capable, with limited debottlenecking of some part, of producing above current capacity.

Licensors determine the fees they will charge for annual capacity or production levels by reference to a scale, on one side of which are increasing increments of 'capacity' and on the other are corresponding paid-up fees. The scale may be linear but will frequently be non-linear so that doubling the capacity will not double the fee. The scale will be applied rigidly or flexibly depending upon the negotiating latitude of the licensor's licensing policy (including any 'equal-treatment' (MFN) clauses in earlier licences). The 'capacity' referred to may be design capacity or achieved annual capacity.

The licensor will ordinarily insist either that enlargement of defined capacity-determining parts of the plant requires consent or that such enlargement will necessitate additional fees. The latter arrangement is better for licensees.

You will appreciate that there is much scope for ingenuity in deciding what should be measured and compared when a plant is enlarged. The reference point might be the former design capacity but, more fairly to the licensee, it should be the actual capacity as measured under defined conditions (as near to normal production conditions as possible) or, failing that, the average annual capacity achieved during the working period of the original plant. The new capacity will have to be a measured quantity, say the value determined by reference to a realistic test run, which again should approximate to normal production conditions. The difference in 'capacity' values will then be used to calculate (or read off a scale) the additional fee required so as to make the enlarged plant fully paid-up.

If the fee the licensee has to pay for 'capacity' is determined not by the assumed design capacity but by what the plant does under test conditions or in the course of actual commercial duty over a long period then again careful definition is required and caution in converting test run performance to annual capability. Plants rarely run for 365 days a year, or continuously. Operating days may not be of 24 hours' duration. New plants do not perform in the same way as well run-in plants; they do better or worse. During year 1 demand for product may be lower than during year 2. Productivity will be lower while 'learning' occurs and so on. Licensors usually ensure they have a significant up-front payment even when achieved annual make is the determinant of fees in the longer term. This payment will be credited against fees due for achieved capacity.

Sometimes, licensors will make receipt of an initial payment a condition precedent to his performance obligations coming in to force or, even, the licence grants taking effect.

If there is no cut-off date for payments or alternatively no cut-off capacity figure, then licensees are liable throughout the royalty payment period to payment of fees for capacity increases achieved by applying their own operating developments as well as those acquired from third parties (who may also demand reward) unless express exclusions are written into the licence agreement.

Some additional points are:

(1) A licensee should ensure that any advance licence fees for paid-up capacity fall due and payable in stages that are milestones along the road to plant beneficial operation. In that time frame (from agreement signature to operations) the project could be aborted or postponed.

(2) Payments for rights of use should be kept separate from payments for design packages or design services. For tax and investment grant purposes they are treated very differently. Additionally, in developing countries, authorisations go through different channels or are subject to different criteria.

(3) It is possible to have running royalty schemes overlaid on paid-up capacity schemes.

(4) The effect of termination of the licence needs careful consideration. When a licence 'terminates' by effluxion of time (usually called 'expiry'), it does not follow that all the parties' obligations come to an end. Rarely is that the case. It is a question of construction of the contract, and it is vital to know what the position is. Termination will bring payment obligations (except for accrued sums) to an end; it will also bring to an end information exchanges and services performance as well as all options for additional licence rights.

At a minimum, the licensee should ensure that he can continue doing what he was doing before expiry and, ideally, what he had the accrued right to do but has not yet implemented. Surprisingly, though, it is known for licensees to accept a shut-out period following expiry of their licences even when they relate to what are plainly central business activities of the licensees.

At termination the licensee may have the right to expand freely or increase production using the same technology and technical information, even if the licence agreement is silent on the point, but only if:
- there are no impeding patents; and
- his obligation not to use the licensor's technical information except as licensed has ceased to apply. Normally, however, the obligation to confine use of licensor's information to duly licensed operations will survive 'termination' of the licence agreement according to the express stipulations in the agreement. The specified shut-out period may be 5 years, 10 years, or indeed until the information comes into the public domain, but in the case of specific collated plant design information that means effectively for ever, so that the licensee cannot build a clone of his licensed plant(s) even though all the principal features of the licensed technology may be publicly known.

ROYALTY LEVELS

There is no entirely satisfactory answer to the question: what is a reasonable royalty to pay for the benefit of practising someone else's developed technology or patented invention? The trite answer is: it all depends.

The price charged for technology is ordinarily determined more by the perceived business value to the licensee than by the cost to the licensor of supply of the technology or, unless the licensor's business is technology development for sale, by the cost to the licensor of development of that technology. A technology development company, for instance, may have to plough back as much as 50% of annual licensing income into continuing R&D programmes. Its sales force, promotional activity, and support overhead staff may easily absorb another 25% of licensing income. These realities have to be factored in to the price charged for technology. In the case of a company whose main business is making and selling a range of goods but which also seeks income from licensing certain of its technologies, it may spend, say, 2-5% of its total turnover on R&D. Turnover is provided by the successes of past R&D and acquisitions. But a significant proportion of R&D expenditure is inevitably consumed by failures. (Indeed, in special fields such as pharmaceuticals R&D costs have to be a much higher proportion of turnover because so few new products are ultimate successes.) Licensees are buying only proven successes. A royalty of 2-5% of net revenue from sales of the product of licensed technology could therefore be said to be a fair price for a licensee to pay and many actual licences do, as it happens, stipulate royalties at that sort of level. There are, however, other factors. A licensor who also makes and sells the licensed product may regard licensing income (after deducting his licensing costs) as pre-tax profits and be very content to receive a royalty of, say, 1-3% of his licensees' net sales income from the licensed product. Again, in fact, many actual licences stipulate such levels of royalty, especially if the product is a bulk commodity or mass-produced item and the licensor has an active licensing programme. Of course, a producer-licensor will have a good feel both for his licensee's production costs and for the likely achievable product sales price in the relevant market. He can, therefore, estimate what his licensee could afford to pay and this may cause him to shade his offer higher or lower. Indeed, if the value of the technology to the licensee is in the nature of production cost saving the royalty levied has to be a proportion of that saving, say one quarter to one third.

Commonly, there will be competitive bidding between licensors possessing broadly similar technologies and then market forces drive achievable licence royalties down, just as prospective intense product competition in the market place will mean there will be smaller returns from which to pay royalties to licensors. All that said, rule-of-thumb benchmarks for royalties might be:
- Commodity product, multi source licensed technology, 1-2% NSI.
- Commodity product, but only two or three licensors, 2-3% NSI.
- Premium bulk product, two or three licensors, 3-5% NSI.
- Specialty premium product/item of equipment, 5-10% NSI, occasionally 10-15% NSI.
- Unique high value product such as a pharmaceutical 20-30% NSI.

Paid-up fees for production technology for a plant producing a commodity or bulk product might be the net present worth of royalties for, say, 6 to 8 years' sales at nameplate plant capacity instead of, say, 10 to 12 years' running royalties on actual sales. If the market for the product and its continuing profitability can be assured this would be a reasonable investment by a licensee.

FOOTNOTE: TECHNOLOGY DEVELOPMENT AGREEMENTS
Technology development agreements are not really licence agreements at all in that the objectives of the parties are utterly different from those of archetypal licensors and licensees. And yet in their provisions they echo the sorts of

considerations that underlie licence agreements. (See Appendix III, Sections 4 and 5.)

The aim of these arrangements is the creation of useful technology by bringing together the complementary knowledge, skills and/or resources of the co-developers.

The intention may be that each party will himself use the developed technology, or that only one will, or that reward will be achieved solely through licensing third parties.

A simple bipartite arrangement may involve one party working to the order of the other, to his targets, on a contract research or development basis. The contractor will be paid for his efforts but will not retain ownership of or control over any inventions, results or patent rights generated by his work. He works as an extension of the R&D function of the principal party but may be entitled, if his contract says so, to some reward if the technology is exploited.

In many cases, the arrangement is much more complex and each situation will be unique. There are no standard forms and no customary practices.

Thus, each party may require rights of use and rights to license others, or at least a right to share revenue from licensing by the other to agreed licensing policies and practices.

Either party may require a right to veto licences for his own backyard and perhaps require exclusivity for manufacture in his own territory. Each party will, perhaps, require reward for its contributions of effort, expense, and creativity should the other party exploit the developed technology by manufacture and sales.

The following are at least some of the issues that must be faced and dealt with explicitly:
- what is the development programme and its split of activities? Who directs and manages the individual programmes or joint team work?
- how are costs and expenses identified and borne?
- what cross-disclosure obligations are there and in what field?
- what confidentiality obligations govern input information from either party to the other, and information and results produced in the development work?
- what rights of use for received input information and for the information and results produced in the development work are allowed?
- what licences, and sub-licensing rights, under either party's existing patents are granted to the other? For what field and on what terms?

- who owns inventions made directly in the course of the development work and who decides on the patent policy? How are patenting costs borne and who does the drafting and prosecution?
- who has the right to grant licences under patents for development inventions? How extensive is the right?
- how does the reward-sharing scheme cope with different inputs of existing information and existing patents, and different contributions of effort, cash and creativity? Is a 'relevance weighting' sensible or practical? If the best the parties will do is 'agree to agree on the day' how do you ensure an outcome and not an impasse? Do you instead go for an arbitrary reward sharing formula at the beginning and simply agree 'in the interests' of both parties to review the equity of that formula when the development is complete. Is there a place for a reference of disagreements to an expert (a 'third-party intervener') or to an arbitration body. How effective, and how final can, or should, the decision be?
- when is the development work to be deemed completed or deemed a failure? When should either party be free to withdraw? What rights should he take with him? What right does the remaining party have to form a new arrangement to continue the development with another of his choosing? If then successful, does the earlier withdrawer have any reward from ultimate exploitation?
- when the development is over, how are patent collisions avoided from independent further improvements made by the developers separately from the same base of knowledge and with years of shared thinking?
- how do the agreed arrangements and their likely effects square up against anti-trust laws, competition laws such as the Rome Treaty, and restrictive trade practices (if relevant)?

An especially challenging type of collaboration agreement is that between a developer of a new material and a developer of a commercial use for that material (or of equipment capable of handling it to convert it into a more readily useable form). At the outset, both have a drive to see the co-operation succeed but, following success, their interests diverge. Neither wants to be tied to the other.

5. THE STRUCTURE AND WORDING OF LICENCE AGREEMENTS

In this chapter the typical contents of a full technology licence involving technical information and patents are described. The ingredients of information licences and patent field licences are also covered, since these latter licences constitute the complementary halves of a full technology licence. Appendix III contains annotated sample agreements, but no attempt is made to offer 'standard' clauses, for two reasons. First, various publications already exist which set out suggested clauses for licensing agreements. They are useful references for legal and patent advisors — if commonly biased towards the licensor's standpoint and heavier on his rights and remedies than on his performance and obligations. Another set with similar or different built-in assumptions would contribute much in length to this work but probably very little to its worthwhile content. Which leads to the second reason. The licensor/licensee relationship is too important to be constrained and distorted by an ill-chosen and poorly understood set of standard terms. The form and content of any licence agreement must be determined solely by what the parties mutually intend their respective obligations, grants, and responsibilities to be in the particular situation before them. They need to approach their negotiations from an informed standpoint and they must understand clearly what they want to say, and what they do not want to say. If that is achieved, it is a relatively easy task to find language to capture their intentions. Then, standard clauses may be used as drafting aids and as checklists. If well-tried language fits the bill, of course use it.

Plain, direct and precise language is always best; there is no need for jargon, although some is now so firmly entrenched that it is difficult to avoid it altogether. Ambiguity is a recipe for later differences at a time when the negotiating strengths of the parties (and their attitudes) may have changed. How often do you hear people say "the spirit of the agreement was such and such"? This is usually an admission that the actual wording of the agreement did not achieve the intended purpose or that it was avoidably ambiguous, but sometimes it can suit the parties to leave possible ambiguities unresolved as in all commercial dealings.

The licensor will usually table the first draft licence agreement. He will call it his 'standard terms' and he will declare that a laundry list of other parties have found it entirely acceptable. A potential licensee should not be reticent about modifying these so-called standard terms to improve their substance or form. There are no 'Brownie points' for conciseness or style at the expense of relevance and clarity. He should take a step back from these standard terms and he may then see that calculated language has been used. The licensor's obligations and grants will be a defined minimum; the licensee's obligations will be captured in umbrella language or by elastic terms that are capable of being given an unduly broad or an unduly constricting meaning, to the disadvantage of the licensee.

Licence Agreements start with a preamble, naming the parties, their registered offices or principle places of business, and the date on which the agreement was reached (although it may take effect on a different date). Then follows the 'Recitals' which set down the basic reasons why the parties are making the agreement, what the objectives are, and perhaps also what the key representations are on which the agreement is founded. The Recitals have a contractual significance and a wider legal relevance. Contractually, they are guides to construction of later substantive clauses in that, if the language of those clauses admits such a construction, they will be read so as to be consistent with the Recitals. The wider legal significance arises when the Recitals state the representations which have induced the licensee to enter into the agreement, such as that the licensor owns the technology and has the right to make the licence grants and technical disclosures promised in the agreement, and that he or his other licensees have operated the same technology successfully at a commercial scale. It is inconceivable that such representations could be untruthfully made as a result of honest mistaken belief. If, therefore, they are untrue, remedies under the law of tort will be available for loss and damage caused as a result of intended reliance on those false representations. Should such key representations be incorporated in the body of the agreement as contractual representations/warranties instead of in the Recitals, then the consequences of their being false will be determined under contract law, plus of course any relevant statutory provisions concerning misrepresentation and fraudulent dealings.

The substantive terms and conditions of the licence agreement are contained in a series of Clauses (or 'Articles' or 'Sections'), each broken down into sub-clauses (or 'paragraphs').

DEFINITIONS CLAUSE

There are three sorts of definition:

(1) Short forms of well known terms, eg day/year = calendar day/year, te = metric ton (tonne), £ = pound sterling.

(2) Convenience terms which already have a general meaning but which are given a slightly different slant or greater specificity for the purpose of the Agreement, eg Subsidiary, Affiliate, Net Selling Price, Force Majeure.

(3) Terms which have no ordinary unique meaning, or can mean different things in different circumstances, and which are used as a shorthand for specific, complex items that are spelled out in detail, eg Patent Rights, Technical Information, Beneficial Operation Date, Effective Date, The Process, The Plant, The Manufacturing Territory, The Sales Territory, The Agreement Field, The Products, The Design/Annual Capacity. Apart from the short forms, defined terms are usually begun with a capital letter, or wholly capitalised, to show they are being used in the defined sense.

Sometimes a defined term is relevant only to one clause or to a few juxtaposed clauses. If so, it may be relegated to that clause or the first of the relevant clauses.

The agreement effect of a defined term is of course only ascertainable from the clauses where that term is used.

INFORMATION SUPPLY CLAUSES

Here will be set out the licensee's entitlement as regards initial technology inputs and the technical exchange arrangements (if any) for future improvements and developments and operating experience.

It is important to specify the types of information, the scope of supply, the depth of treatment and the form in which the information is to be supplied. For the initial input from the licensor a basic process design package will be an essential requirement. Its contents must be specified (usually the details are put in a Schedule) but timing and reviews are also important topics. The package will stop short of the sort of engineering detail which the licensee, or his contractor, can carry out himself. Some stipulation ('warranty') that the package will be sufficient to enable the licensee to complete the engineering design, construct and operate the prospective plant to produce the intended quantity and quality of products is needed. Relevant definitions will feature prominently, eg Plant, Process, Technical Information, Product(s).

If the licence is a production capacity licence and not just a one-plant licence, the licensor's initial inputs will not necessarily be any more comprehensive but, clearly, a licensee would wish to learn more of the philosophy, the reasons for elections, the fundamental science and the design correlations. If so, these extra requirements must be specified; they will not be volunteered by the licensor.

Provisions on future exchange of information need to set down the class of information that qualifies for exchange, the time frame of qualifying information, the period of time over which the parties can call for and discuss the information to which they are entitled, and the mechanism for a workable and effective exchange. An undertaking to 'make available' information is not the same as an undertaking to 'supply'. A supply obligation without a time frame, or exchange procedures, is a recipe for minimal useful exchange unless both parties are fully committed to the exchange as being in their interests.

Apart from the general exchange of future improvements and developments, there is the question of the licensee's requirement to be able to call on the licensor for process design help for any licensed modification or enlargement of the agreement plant or for any further licensed plant. Inevitably, such a requirement must be expressed rather generally and many matters will have to be left for agreement and decision when the time comes for the design work to be done.

It is worth stating here that agreements are not just about obligations that have sufficient specificity to be legally capable of being enforced by an order for specific performance (most unlikely though for an activity such as process design) or giving rise to damages for breach (not really much help either if what is really needed is a design which only the licensor can supply). Agreements are also about undertakings that it would be difficult for an honourable company (or one sensitive to its public image and reputation as a licensor) to walk away from without very good objective reasons. In this writer's experience, when licensors have given a plain commitment they have honoured it even though perhaps they could not have been forced to do so or been 'punished' for failure.

SERVICES SUPPLY CLAUSES

The sorts of things dealt with here are licensor supervision or scrutiny of detailed engineering design and critical items of plant equipment at vendors' works (usually for process conformity and consistency with the basic design package), licensor presence at key stages of plant construction and at completion, at start-up, and perhaps later if problems should arise. Presence during any guarantee test runs is usually dealt with in the guarantee clauses/test run procedures. Licensee's needs for operator training should be covered here as might, in an extreme case, hands-on running or supervision of the licensee's plant by licensor's staff for an initial period of operations.

It is sometimes simpler administratively to deal with specific design needs and services needs for a first plant in a separate Design/Services Agreement(s). If that is done, then the licence agreement must suitably cross-reference the Design/Services Agreement(s) in respect of, for example, rights to use supplied information, and the effects of design errors on Plant performance and consequent liability for failure to meet performance guarantees. Legal advisors must also take particular care to ensure that if the party(ies) to the Design/Services Agreement(s) should be Contractor(s) and not the licensor that overall responsibilities and liabilities are clear and do not get stranded in a legal no-mans-land between distinct self-contained contracts.

THE GRANTS AND GRANT-BACKS

These are core provisions, and in earlier chapters the principles and problems have been discussed in some detail.

The Grants Clauses must determine in clear terms:
- what manufacturing and trading activities the licensee may (patents apart) use supplied licensor information for. (Relevant definitions: Technical Information, Plant, Process, Products, Agreement Field.) The Secrecy Clause will usually be cross-referenced;
- what facility or plant is the use to be confined to, or, if a production capacity licence, what territory is use to be confined to as regards manufacture. (Relevant definitions: Plant, Manufacturing Territory, Products.) Again, there may be a cross-reference to the Secrecy Clause;
- what contractual restriction on sales is there (patents apart). (In many cases, such contractual restrictions will not be allowable or enforceable, nor will they be acceptable to the licensee);

- what patent rights are granted and/or what immunities from suit under licensor's patents are agreed. (Relevant definitions: Patent Rights, Manufacturing Territory, Sales Territory, Process, Plant, Products, Agreement Field.) This has two aspects: what Patent rights are the subject of the grant, and what activities are 'licensed', or enjoy an immunity from suit under the defined Patent Rights. Deal with manufacture and sales (including sales into foreign markets) specifically. Deal with patent rights for important raw materials, catalysts, consumables, equipment, processing aids even though information supply may exclude details on the production of these. Consider patent rights for important end-uses even though, again, information supply may not extend into these areas.

Grant-back Clauses must specify:
- what rights are granted for use of future exchanged information;
- what rights are granted under patents for inventions contained in or exemplified by future exchanged information which would otherwise be infringed by their use in the Plant, Process, Products, Agreement Field, Manufacturing Territory and Sales Territory or, alternatively, what rights are granted under patents which relate to the Process, Products, Agreement Field and which are obtained in a specified future period and existing in the Manufacturing Territory or Sales Territory.

Because of the crucial importance of these Grant/Grant-back Clauses, and the complexities of patents, a more detailed analysis and the pros and cons of different wordings are set out in Appendix II.

The Grant and Grant-back Clauses must also set out:
- what manufacturing sublicensing rights are granted, if any. (Relevant definitions: Subsidiaries/Affiliates, Patent Rights, Manufacturing Territory, Process, Products, Agreement Field);
- what sublicensed rights (if any) or hold-harmless undertakings customers for Products enjoy in the area of downstream uses.

THE PAYMENTS CLAUSES

The relevant Issues have already been discussed fully in Chapter 4. Essential points are:
- what activities/products qualify for royalties or fees?
- how are the royalties/fees calculated?
- when are royalties/fees due?
- when are they payable?
- how are they reported and paid?

- what escalation, what credits, what maxima or minima, what auditing?
- how long does the obligation to account for royalties last?
- tax questions, ie withholding taxes, VAT;
- are royalties in respect of patent licences separately identified from royalties for use of confidential information? If not, does it matter?
- are fees for designs/services excluded from royalties/fees for rights of use? If not, does it matter?

PERFORMANCE GUARANTEES CLAUSE

In theory, the concept of performance guarantees is fine; in practice they are not usually ever fully satisfactory to a licensee. What they often do is merely exonerate a shoddy job or even a fiasco. Their legal effect is often merely to limit the liability of the licensor for breach of contract. It is licensors who want guarantees because they can ensure that whatever happens they walk away with money in hand, or an expectation of royalties, sufficient to offset licensing costs and usually to contribute a significant level of profit on the transaction.

The essential elements of a guarantee clause are:
- definition of guaranteed parameters;
- specification of guaranteed levels for the parameters;
- specification of Test Run Conditions, ie exactly how the plant or facility is to be run during the Test Run(s), what is to be measured and how often, how long the Test Run is to continue, what happens if the Test Run is interrupted, what is to happen if the first Test Run is a failure, what changes to plant or operations can be insisted on by the licensor, how many shots are allowed to the licensor for achieving guaranteed levels or bettering plant performance, who pays for plant mods, under what circumstances can the licensor walk away, what actions or inactions by the licensee forfeit his right to a Test Run, a second or third Test Run, and all so-called liquidated damages (meaning the sums payable by the licensor to the licensee for the plant's failure to meet guaranteed performance);

- specifying the liquidated damages that are payable for measured departures from guaranteed levels for the different guaranteed parameters. (Sometimes a credit system is included so that better than guaranteed performance in some respects can offset, at least in part, poor performance in other respects);
- specifying the ceiling on liquidated damages (inclusive of, or after crediting, as the case may be, costs for plant mods).

(Note: The licensor will also insist that the licence agreement specify a contractual limit on his maximum aggregate liability under all legal heads in the licence agreement as well as setting a limit on his liability for these liquidated damages.)

It should be borne in mind that failure to meet guaranteed levels would not necessarily, in the absence of the guarantee clause, have implied a breach of contract. Liquidated damages are simply payment adjustments agreed to as a contractual matter in certain defined circumstances. They need neither fault on the part of the licensor nor actual damage to the licensee to support them. The name 'liquidated damages' is really a misnomer because the levels stipulated can seldom be regarded as a genuine pre-estimate of the probable direct loss or damage to the licensee arising from material departures from expected performance. 'Limited damages' would be more accurate.

Cases are known of licensees choosing to waive the right to have a test run because it would cost more in lost production, and interferences, or actual added costs than could ever be recovered in liquidated damages. The negotiation of guarantees and damages does, of course, enable the licensee to measure the licensor's confidence in his technology and in his ability to transfer it to the licensee effectively. There is no doubt, too, that the existence of a significant risk to his royalties will concentrate the mind of the licensor; no-one readily gives up say 20 – 40% of his expected income.

SECRECY CLAUSE

This clause deals with non-disclosure of licensor's information and specific categories of permitted disclosures. It will also contain user restrictions to supplement and re-enforce, or even to capture in a proper legal context, the intention of the permissive 'licence to use information' set out in the Grants Clause.

The essential requirements for this clause have already been discussed in Chapters 3 and 4. Appendix II is also relevant.

There are four additional points that should be made:

(1) The licensee cannot now reasonably expect a licensor to agree to independent subsequent developments of similar technology being excluded from the obligations of this clause. This particular gateway should be closed as the price to be paid for the benefits of the licence.

(2) When information relevant to product sales is provided by the licensor, it is sensible to define a category of information ('Sales Information') which is excluded from secrecy obligations so that the licensee can pass it freely to his customers.

(3) The licensee should resist any attempt by a licensor to bind a licensee not to disclose his own information of any kind to third parties, such as, especially, improvements and developments of the licensor's technology. The fact that the licensee would risk disclosure of information received from the licensor if he were to disclose improvements or developments (and thereby breach his undertakings in respect of that received information) is not a sufficient justification for a specific disclosure ban.

(4) US companies are enamoured by a bit of 'boiler-plate' which says that "a combination of integers is not to be regarded as being in the recipient's possession or in the public domain just because the individual integers are". So far so good perhaps but the provision frequently goes on to say "unless the combination itself, and its principles of operation, are in the public domain or recipient's possession". This is in part tautology and in part excessive. Many a combination is in the public domain and can be fully used and copied without anyone really understanding why or how it works. This is an unnecessary addition and still leaves open the question whether combinations whose principles of operation are not understood are in or out. Does a combination, before it is exempted, have to be wholly in the public domain or wholly in the recipients possession? Many an obvious combination straddles both categories.

VALIDATION, TERM AND TERMINATION CLAUSES

Ordinarily, the licence agreement will come into full force and effect upon execution by both the parties, but this need not be the case. There may be, for example, a requirement that the licensor or licensee obtain from his Government departments and Central Bank approval of the terms of the intended agreement. The licence agreement will in such a case say that the agreement shall come into force on a defined date subsequent (the 'Effective Date') being the date when all necessary Government approvals to enter into the agreement so as to be validly and legally bound by its terms and to make the stipulated payments have been received. Sometimes licensors will make their obligation to perform (and the grant of licences) dependent upon receipt of an initial lump-sum payment.

It is usual for a licence agreement to have a given life at the end of which all obligations expire except those expressly stated to have a longer life (and which are kept alive) and apart, of course, from any obligations that have expired already as a result of full performance or express time limits and calendar cut-offs. Specific time limits are common for obligations such as secrecy, use-restrictions, royalties, and supply of future technical information or the grant of licences under future patents, but others such as options for additional licences, and plant design services may be ended by a blanket termination clause. If relevant patents continue in force after the Agreement term, then, assuming an ongoing licence will be needed, the patent licences should be expressly kept alive.

Commonly, the Agreement will say that it will terminate on the later of: a specified anniversary of the date of the agreement, and the expiry of the last to expire of licensor's patent rights; but if there are no express time limits for key obligations and there is a period of future patent rights exchange the termination date is then uncertain and the effect of this on other aspects of the Agreement must be assessed. A licensee must ensure that on termination he knows precisely where he stands on continued right to work under any of licensor's remaining patents in the area of the licence grants which in fact he was working under prior to termination. A clear position on continued use of technical information is also vital because restrictions in this area are usually expressly continued beyond formal termination of the licence agreement by effluxion of time (usually called 'expiry' to distinguish from termination at the election of the licensee or because of his breach of a material condition).

If the agreement allows the licensee to terminate the agreement at his choice before the normal expiry date, special provisions for the post termination phase will be needed. The licensor will insist on a contractual right to terminate the agreement and cancel all 'licences' in the event of unremedied licensee default (see below).

THE GENERAL CLAUSES
These deal with:
(a) Force Majeure
(b) Default
(c) MFN ('Most Favoured Nation' or 'Equal-treatment' clauses)
(d) Unlicensed infringements

(e) Third Party patents
(f) Assignment
(g) Governing Law
(h) Arbitration

(a) FORCE MAJEURE

Any agreement involving future performance obligations of uncertain timing and quantity, and which are liable to be hindered or prevented by events or circumstances not under the control of either party should include a Force Majeure clause to excuse non-performance because of such events and circumstances. Performance should be excused only while there is hindrance or prevention.

In a straightforward licence arrangement there are not many performance obligations that are outside the parties' control but the clause is a useful contingency provision. Sometimes licensors will insist that the clause should not excuse non-payment of fees and royalties through Government or Central Bank intervention.

Political risks of this sort are of course a hazard but a fairer way to deal with the problem might be to require early payments and to allow the licensor to suspend his performance if payments are withheld.

(b) DEFAULT

Licensors always insist on a right to terminate the Agreement (and to revoke all licences as a sequence) if the licensee makes any breach of the contract. Strictly this is excessive and potentially harsh in its effects because a right of termination for all breaches is not something the law would ever admit, in the absence of an express contractual right agreed to by the parties. Even then enforcement may be difficult or impossible in a foreign jurisdiction. In practice, with a Force Majeure clause also present, there is probably not much to get excited about. Licensees ought to be exposed to the risk of withdrawal of their licence rights if they fail to safeguard the licensor's secret and proprietary information or fail to account properly for royalties and fees. Needless to say, licensors are likely to be reasonable and circumspect over exercising their discretion to terminate.

Sometimes licensors will insist that liquidation of the licensee company, or commencement of winding up proceedings (but not merely reconstruction) will be deemed a breach and give the licensor an immediate right to terminate. National laws in the licensee's country can frustrate any such attempts to terminate.

THE STRUCTURE AND WORDING OF LICENCE AGREEMENTS

(c) MFN ('MOST FAVOURED NATION' OR 'EQUAL-TREATMENT' CLAUSES)

These clauses are not the benefit to a licensee which they might at first sight appear. In these clauses, licensors agree that if they should subsequently grant a licence of substantially similar scope to another party in the relevant territory on more favourable terms the licensor will offer those more favourable terms to the licensee, subject always to the licensee accepting any less favourable terms. Sounds fine, but what is 'substantially similar scope'? What are 'more favourable terms'? How will the licensee know if the licensor chooses not to tell him? Licensors like these clauses because they give them an easy pretext for not reducing their offered licence payment terms or revising other onerous clauses but, at the same time, if they should wish to do so to get a particular piece of business they usually seem to be able to negotiate some agreement change without activating the MFN clauses of their other licences!

(d) UNLICENSED INFRINGEMENTS

An exclusive licensor under some jurisdictions has the legal right to sue patent infringers. A non-exclusive licensee may be given that right in the licence agreement but it is not common to do so. As a minimum a non-exclusive licensee does not want to be obliged to pay royalties if others are allowed to infringe and do not have such a charge on their business. A licensee will have a struggle to get a licensor to agree to sue all infringers. He will probably have to be content with a stipulated right to withhold some or all royalties in the event that there is infringing manufacture or sales in his Territory. Perhaps a 'significance criterion' will be stipulated. Perhaps, withheld royalties will become payable once infringement ceases. Certainly future operations, after infringement has ceased, will once again involve royalty payments. There are no easy formulae; the parties must make the best bargain they can.

(e) THIRD PARTY PATENTS

This, too, is a subject of some complexity and the licensee must negotiate the best deal he can. No licensor ever unreservedly warrants that third party patents will not be infringed by operating the Process in the Plant in the Territory, using licensor's Technical Information because he has no control over the licensee's operations. The best a licensee can hope for is that the licensor will indemnify him for costs and damages, or settlements, or additional licence royalties to the third party patentee if the infringement derives directly from use of information supplied by the licensor in ways intended or for the intended purposes. A licensor will usually insist on the limit of his financial obligations to the licensee being some proportion of royalties paid or payable in future by the licensee. The proportion might be 30 to 50% but subject to, say, a 50% limit on all liabilities under the licence agreement (ie including liquidated damages for failure to meet performance guarantees). Plainly, this could fall far short of full indemnity especially if production is halted by injunction and so a licensee must at the earliest practicable time do a patent search and seek to identify any problem patents and use his own initiative to secure a tolerable outcome before incurring major investment and business costs.

(f) ASSIGNMENT

It is valuable to a licensee to have an express contractual right to assign the licence within a larger group of companies of which the licensee is part. This would allow corporate restructuring. Similarly, a licensee may seek a right to assign the licence to any successor to the relevant business who acquires the licensee's manufacturing assets including the licensed plant. It is important that the assignment should be notified to the licensor and that it should take effect only when the assignee has agreed in writing to be bound by all the terms and conditions of the licence.

Plainly, such assignment rights require modification of the non-disclosure obligations to permit disclosures as necessary to potential assignees to enable them to assess what it is that is being assigned. Also, necessary concomitant disclosures to Government agencies and advisors should be provided for. Licensors, for their part, can be understandably reluctant to allow an unchallengeable right to assign the licence to anyone. They will often insist that any assignment shall require consent so as to be able to prevent the licensed technology passing to a major competitor or to a licensor of competitive technology.

Of course, upon any assignment the assignor (the initial licensee) should continue to be bound by strict secrecy obligations and there should be a total prohibition on his further use of licensor's information.

(g) GOVERNING LAW

The choice of governing law is usually made by the licensor and he will choose his home jurisdiction (which he knows) and rely on enforcement in the licensee's country under ruling practices for enforcement of foreign judgements, or he will choose the licensee's jurisdiction because that is the place of licensee's performance (and where he has assets) and, procedurally, it might be easier to get quick enforcement. In some cases, the licensee's home law will have to apply by Government ruling. It is usual to stipulate that validity, construction and performance will all be governed by one chosen law but there is no reason why this must be so. Plainly this is a complex area and proper legal advice is needed.

Licensors and licensees hope, of course, that they will not have major disagreements and that if there should be a dispute they will resolve it amicably. If they cannot, they can always agree on the day to seek expert determination of a disputed point, or arbitration if there is a major interpretation issue. Very commonly, the parties will agree in the licence agreement not to have recourse to the courts of law to resolve any disputes that might arise but to seek arbitration.

(h) ARBITRATION

There are many eminent arbitration bodies who have well established procedures. A common provision in licences is one giving either party the right to seek final determination of a dispute by arbitration under the rules of conciliation and arbitration of the International Chamber of Commerce, if a dispute is not settled by amicable discussion in a reasonable period of time. A forum should be specified and commonly this is the commercial centre of a neutral territory. If the parties want to appoint their own arbitrators and to stipulate their own procedural rules for the arbitration process, then they must take professional advice at an early stage because there are many potential hazards which can frustrate the party who wants a quick final decision, even with the best drafted clauses.

Indeed, some national laws will allow appeal to the courts of law from arbitrators' decisions even though the licence agreement said the arbitration was to be the final determinant. So-called 'special case' procedures are known also under which there is a right to have a preliminary point of law decided by the courts. UK law now allows the parties to international licence agreements to exclude recourse to courts of law and for domestic licences the parties will be bound by an agreement that an arbitrator's decision shall be final if the agreement to that effect is made after the dispute has arisen.

An arbitration clause will often make any dispute "arising out of or in connection with the agreement" referable to the nominated arbitration body (but see below).

If enforcement of an arbitration award is necessary, this too will involve the regular courts.

The problem with arbitration and with court proceedings is that under UK (and many other) systems, the arbitrator or the judge has no power to resolve genuine ambiguities by imposing a fair solution or by filling gaps in the contractual provisions. In a few countries the courts and arbitrators have such powers conferred by statute or they assume such powers under so-called inherent jurisdiction.

Whether arbitration or court proceedings are cheaper or better is a debate on which practitioners fall into two polarised camps, often determined by the latest good or bad experience! The best that can be said is: it all depends.

The more technical or performance-related the issue the more arbitration might seem better. Similarly, a patent interpretation issue might be more expeditiously handled by a suitably qualified expert or arbitration panel. The parties are of course at liberty to specify which sorts of issue shall be referable to an expert or arbitration body and, conversely, which shall not.

APPENDIX I — INTELLECTUAL PROPERTY RIGHTS

This appendix includes some further information and comment on Patents, Registered Designs and the new UK Design Right which the reader may find helpful but which would have made Chapter 1 unduly heavy going.

PATENTS

In Chapter 1 I said "...delineation of the sought-for monopoly in the form of patent claims is the key determinant of the maximum monopoly rights granted to the patentee". The drafting of patent claims, and the analysis of the invention that precedes it, are activities which the qualified patent agent or attorney would regard as his special province — where he really does earn his salary or fees!

The function of the claims is not to describe the invention (although they may incidentally do so), nor is it to promote the merits of the invention. It is to distinguish the subject matter for which a monopoly is sought from the prior art by defining those elements that, collectively, are the essential ingredients.

The patent claims are arrived at by an intellectual process of regressional analysis on what has been made or done by the inventor to see what the essential new factor at work is believed to be. It is a search for fundamentals and a labelling of all else as irrelevant, or merely convenient, or just preferred. The patentee then claims all systems, compounds, articles, machines, methods, etc (as the case may be) of a relevant general class that are characterised by possessing that essential factor (expressed as, say, a particular order or arrangement of steps or stages, a particular class of additive or substituent, a particular attachment or arrangement of working parts, a particular regime of operating conditions, or a generic chemical- or composition formula). This is his main claim. The patentee also claims in a hierarchy of subordinate claims of reducing scope the subject matter of his main claim further characterised by the inclusion of the preferred and convenient additional factors, singly and in combination, which emerged from his regressional analysis. These are separate sub-monopolies each of which will stand or fall on its merits if the patent is rigourously challenged. Their drafting is equally as important as that of the main claim. Indeed the framing a defensible generalisation between a main claim that is struck down and the most preferred embodiments or fields of application may be crucial to a worthwhile monopoly. The commonest reason for invalidity of patent claims is that the fundamental principle which the inventor thought he was the first to discover is in fact already in the public domain along, possibly, with some or many of his preferred and convenient embodiments. The invention which in fact he made (viewed as an addition to what already exists) may be way down his hierarchy of claims in the patent as granted. Another common reason for partial invalidity is that an earlier author or practitioner, without disclosing to the public the unifying theory that this patentee has contributed for the first time, has nevertheless taught the public a particular sub-class or even an isolated example of what the patentee has claimed.

No erudite theory can save a claim that covers something already known even though it might earn a Nobel Prize. The most common other reason for invalidity (at least of the broadest claims) is that they are not sufficiently and fairly supported by the core factual patent disclosure when all pure speculation, bad theory, and circular reasoning is stripped away. Even so, most patents give effective cover for much more than a mere duplication of the specifically described embodiments and worked examples.

Basically something that is not caught by the language of a claim is not an infringement of it even though, given more vision or better advice, a valid claim covering that thing might have been drafted. Claims are not literal straight-jackets however; they must be construed in the context of the art in which they fall, with special technical terms given their appropriate contextual meanings, and also be interpreted in the light of the description and examples in the body of the patent specification. Courts in the UK and USA, for example, have in a few celebrated cases adopted this process of construction of claims and declared activities not caught by the literal, dictionary meaning of a claim to be infringements. The justification was either that some critical word did not here bear its narrow dictionary mathematical or scientific meaning (eg 'vertical' or 'co-axial'), or that some element of the claim was clearly inessential and all the infringer had done was substitute a mere obvious equivalent doing the same job in the same way for the same purpose. But, that said, an element clearly made essential by the specification and its claims will be adjudged essential. Something that merely adds a further improving element, however meritorious and patentable in its own right, does not escape infringement.

REGISTERED DESIGNS

Even before the recent amendment to the UK Registered Designs Act when, for the first time, a requirement of relevant aesthetic appeal was introduced this form of protection for designs applied to industrially exploited products was little used. The procedures and fees were not, and still are not, onerous. The reason was, perhaps, that infringement was perceived as likely to result from copying, and the automatic similar protection then available under copyright law was felt to be adequate. Indeed, the remedies available under Copyright law were a major deterrent to would-be infringers. Now, the Registered Designs Act and the Copyright law in this area have been fundamentally changed but the new Design Right (see below) has been introduced. It is hard to predict what the net result of this change will be, but the protection available under the Registered Designs route will remain an important factor in technology protection for singular designs of great public appeal.

To be registrable, the design which is applied to articles in respect of which registration is sought must be 'new' as judged by its appeal to the eye. In assessing whether a design is new, all existing applied designs are taken into consideration and not merely those applied to articles of the class for which the registration is sought. Immaterial differences and common trade variants are disregarded. With a few exceptions and some legal 'deeming' in a few special circumstances, the expression 'new' has its ordinary dictionary meaning. The Register of Designs is open to the public.

The protection granted to the proprietor of the design (usually the designer's employer) is a right to stop others making or importing or supplying for use articles of the kind for which the design is registered to which the same design has been applied (or moulds or patterns by which they may be made). As for patents, this is a monopoly right so that there is no need to show an infringer has copied the design. In practice infringement will usually derive from conscious or unconscious copying. Indeed, the new requirement for registrability that aesthetic appeal must be important is more likely to mean that infringers will be copiers. The term of protection has now been raised from 15 years to 25 years from registration (subject to payment of fees). The infringing articles need not be identical in design to the registered design but differences must not be material if an action for infringement is to succeed. Use of infringing articles is not preventable (unlike with patents) and the copying of principles of construction or operation evident from an inspection of an article is not an infringement of any rights granted by design registration for that article.

Plainly these rights are narrow rights and since technology rarely expresses itself in a single and unique aesthetic form they are likely to be relevant to technology licensing only in special cases of, say, consumer goods and appliances where elegance of design can be as important as or more important than superiority of function to the consuming public.

THE UNREGISTERED DESIGN RIGHT
The main features of this new form of protection for designs applied industrially to marketed products are:
- it applies to aspects of shape and configuration of the whole or part of an article;
- the protection is automatic on creation of the design and its fixing in some drawing, writing, or article made to the design;
- no registration is needed before rights can be asserted;
- it applies to functional as well as to aesthetic articles (although must fit/must match aspects and surface decoration are to be disregarded);
- there is no concern with appeal to the eye or artistic merit;
- the design need not be new as such but in its application to the product it must be an original work and it must not be a commonplace design;
- the term of protection is 15 years from the design's creation or, if shorter, 10 years from first marketing anywhere (not unduly generous). Additionally, licences of right are available during the final 5 years;
- the protection is available to EEC firms and individuals and to those foreign nationals and firms whose governments provide similar reciprocal rights for UK-originating designs. Where, on those qualifying rules, a design right is denied, an exclusive licensee who is first to market articles to the relevant design in the EEC or a reciprocating foreign country obtains a Design Right. (It is therefore to be hoped that this type of design right becomes universally popular);
- although infringers must be shown to be dealing in copies of the design (knowingly, if an importer, wholesaler or retailer/hirer) this is not likely to frustrate assertion of the Design Right where the design has distinctiveness and characteristic features that are not commonplace or merely dictated by must fit/must match considerations;
- the new Design Right does not have retrospective effect. Existing designs, if Registered Designs, continue to enjoy the protection they had before the recent amendments to the Registered Design Act (eg a 15 year term, not 25 years). Existing industrially applied designs that are not Registered Designs, but could

have been registered, continue for 10 years to enjoy some protection under the Copyright route deriving from copyright in their founding original drawings.

EUROPEAN COMPETITION LAW: HOW IT IS STRUCTURED

The practical impact of Article 85 of the Rome Treaty (the principal Article that influences licensing touching on the European Common Market) has through the efforts and influence of the executive arm, the European Commission, become increasingly legalistic and interventionist. We have moved from a proscriptive law seemingly aimed at self-evident wrongs (with the European Commission acting as counsellor and assessor in the public cause) to a system of direct regulation by the European Commission. This came about in the following way.

Article 85 said agreements affecting intra-market trade and having certain objects or effects were bad but, if they had particular redeeming features, they might be excused. Subsequent Regulations by the Council of Ministers said that only the Commission was competent to decide if a prima facie bad agreement had exonerating features and that all such agreements had to be notified. A limited set of solely national agreements and 'minor' agreements were excused the notification procedures.

Additionally, the Commission were empowered to investigate agreements and to stipulate by Regulation which provisions in certain bipartite Agreements (note 'provisions', not overall 'agreements') may and may not be included in the Agreement if it is to benefit from executive exemption. The system of Block Exemption was then introduced in a climate suggesting that many, if not most, licensing agreements could affect trade and have undesirable objects or effects. The Block Exemptions for patents and know-how, amongst others, that were issued by Regulation, after much consultation and opportunity by interested parties to comment, specified that for licence agreements of the sorts to which they applied certain provisions were excusable, others were harmless, but those in a third category would disqualify the agreement from the benefit of the Block Exemptions. Of course, given the way in which the powers of the Commission have evolved, these Block Exemptions are very worthy instruments affording some comfort to industry and avoiding much needless bureaucratic scrutiny. However, their limitations and adverse consequences cannot be glossed over. Firstly, by their very nature, they must be 'fail-safe'. (It is not the purpose of the Block Exemption to let an appreciable number of agreements slip through that are likely to operate in a way as to offend against

Article 85. Nor is it in the interests of the Commission to abandon its wide investigative and regulatory powers). Secondly, there is an inherent problem of a logical mismatch between provisions in agreements as such and effects on trade between member states and competition within the Common Market, which is what Article 85 of the Rome Treaty is about. We can all sympathise with the desire to translate the general and admittedly loose proscription of Article 85 into a set of manageable rules of precise focus, but there is a price to be paid. The rules we now have tend to reflect an unshakeable conviction that a competitive free-for-all today without regard for the longer term health of industry within the Common Market as a whole, or the ability of Common Market industry to compete against the other major world powers, is a good thing. On a less lofty plane, the rules also feed the fear in industry that agreements that cannot be fitted in to a Block Exemption must be bad. This is of course not the case. An agreement that benefits from Block Exemption may nevertheless be bad and, much more frequently, an agreement that cannot benefit may nevertheless be harmless either because it is not caught by the proscription of Article 85 at all or because the Commission would be obliged following mandatory notification to exonerate it. While there is much philosophical similarity between US Antitrust law and European Competition law (and the US Antitrust enforcement agencies have formidable powers of investigation and prosecution) there is less blanket regulation in the USA and, because there is but one Government involved, a greater willingness to use political power to influence dynamically the working of Antitrust laws in the national interest as perceived by the incumbent administration.

APPENDIX II — THE LICENCE GRANTS

RESTRICTIONS AND FREEDOMS
In this case study, a licensee is investing in a new plant to use a licensor's production technology. The Licensor supplies a front-end engineering package comprising:
- raw materials requirements and specifications;
- general plant layout, equipment descriptions, duties and specifications plus data sheets for main plant items;
- process flowsheets, process descriptions and outline operating, control and safety information;
- product descriptions and specifications.

This information relates solely to requirements for a particular plant (the Plant, if a unit licence, or a first Licensed Plant, if a production capacity licence). From this information, the licensee, usually with the services of engineering contractors and construction contractors, will complete the detailed engineering design for the plant, procure all equipment and other plant items, build and operate the plant. The licensor will carry out reviews and checks and will be on hand at commissioning and start-up as well as during any performance test runs.

The supply of licensor's technical information is purposive in two senses. It is an instruction kit for a specific investment and it is put at the disposal of the licensee only for licensed purposes.

CONTRACTUAL RESTRICTIONS AND FREEDOMS
It is common for licence agreements to include an early prominent statement of the uses to which licensor's technical information may be put in addition to provisions that appear later (usually under the Secrecy clause interwoven with the non-disclosure obligations) which constitute specific restrictions on use. Such statements range from neutral expressions such as "Licensor's Technical Information is available to Licensee for use to practise the Process [in the Plant] [in the Territory] and for the use and sale of Products worldwide", through to

permissive language redolent of patent licensing such as "Licensor hereby grants to Licensee a non-exclusive non-transferable licence to use Technical Information for, but only for, the practice of the Process, etc".

The merits of such statements as an expression of the purpose of the licence arrangements are obvious. Legally, they are no more than statements of consent by licensors to the information they will be supplying being used in whole or in part by the licensees for the recited purposes. The consent is limited to those purposes expressly or by implication.

Although often expressed as a 'licence' or 'permission', the licensee strictly needs neither. Such a statement creates an expectation that somewhere in the licence agreement there will be a qualified undertaking by the licensee to confine his use of supplied technical information to the recited field of use, but this statement is not of itself such an undertaking, and it is unsafe to rely on such statements as necessarily implying collateral undertakings by the licensees should they ill-advisedly be unaccompanied by explicit undertakings.

The explicit undertakings are the legally vital provisions that capture the restrictions on use of licensor's technical information that are created contractually. The undertakings may refer to the earlier statement of consent and so take the form of a simple undertaking not to use licensor's technical information for any purpose other than "as licensed herein" or "as recited in clause...". The vital extra requirement is that the licence agreement must stipulate that these undertakings shall not in any way restrict or prevent the licensee from using any information, however similar, that is available to him from sources independent of this licensed supply of technical information, eg from published sources, prior development or third party sources, all as more fully discussed in the main text.

This is not legal pedantry. The fact that, under a common law duty of confidentiality, these exclusions would be presumed does not mean that they will be implied in a contract so as to cut back on clear agreed broader restrictions. A person can contract to curtail his freedom to use his own or others' information and he can bind himself to make payments if he uses information of a given subject matter class in certain fields irrespective of where he got that information from.

It should not be imagined that such 'broad' restrictions and obligations will not be enforceable. Indeed, it is common for licence agreements to make the payment of royalties or licence fees a strict obligation. The royalties may, for example, simply be a levy on output from the plant that was built to the licensor's design package. Royalty payments are often visualised by licensors as a financial arrangement for spreading payments for the benefit of the licensee in lieu of an initial 'equivalent' lump sum, and not as compensation for a continuing licence to use information while it does in fact continue to be confidential. As discussed in the main text, it is important to lay down clear rules for the position should another licensor's version of 'the Process' be used subsequently for a second plant. There can rarely be any justification at all for a first licensor insisting on payment to him if the second operations do not require and do not involve use of any technical information that is available to the licensee solely by virtue of the first licensor's supply. Equally, the fact that the first licence agreement may have granted rights under the first licensor's patents in terms broad enough to permit the licensee, if he wishes, to work these patents in the second licensed operations is no justification for an 'option' levy. Only if the second operations fall within the scope of the first licensor's patents should the licensee have to account to the first licensor.

Returning to the case study and the narrow issue of allowed use of a licensor's supplied technical information, the agreement position will typically be that the licensee may use each and every item of supplied technical information for the following purposes (and shall not use any such items for other purposes if they cannot benefit from the contractually specified exclusions): to operate a defined class of process (of which the particular detailed process to be used initially is one example) to produce a product of a defined class (of which the intended initial product is an example) in a particular production facility either at a particular location (in the case of a unit licence) or in production facilities located in a particular geographical area (in the case of a production capacity licence). Relevant definitions will be 'Process', 'Plant', 'Product(s)', 'Territory', 'Added Capacity'.

(It is assumed in this case study that there are no contractual restrictions on sales to any territories or markets, whether or not the patent licence is limited so as to make exports of Products actionable as infringements.)

And so, patents apart, the licensee is at liberty, if he knows how, to modify and improve the plant and the technology it uses and to increase production capacity and to use the knowledge and experience he acquires in the course of plant operations for any other business purpose, provided only that he does not in so doing use (or disclose) the specific information he received from the licensor that is covered by the contractual restrictions other than for the contractually agreed licensed purposes and other than on the agreed terms and conditions as to secrecy, payments, etc.

Again, it cannot be stressed too strongly that the position of the licensee at expiry of the licence agreement or following any earlier termination under specific contractual avenues must be clearly spelled out. Licensees should be especially wary of post-expiry contractual bans on continued use of licensed technical information as previously licensed, whatever the position on patent licences may be.

RESTRICTIONS AND FREEDOMS UNDER PATENTS

The purpose of the patent licences is to permit the licensee to do certain things (the licensed acts) which, but for the licence, would or might be infringements of patents owned or controlled by the licensor.

From the standpoint of the licensee's right to do the licensed acts, it does not matter if the licensor has a lot of, some, or no relevant patents. If any particular licensed act is in fact not covered by any patent of the licensor, the licensee does not need a licence but he is not embarrassed by the purported grant of one. The patent licence does not require the licensee to confine his activities to 'licensed acts'. If the licensee does perform 'unlicensed acts' he does so at his peril should the licensor have patents assertable against such other acts.

The licensee is in the same position, no worse, than pure third parties. While mis-use of confidential information in breach of contract terms would create a risk of termination of the licence by the licensor, infringement of patents by unlicensed acts does not unless the licence agreement expressly makes it so.

In defining the licensed acts (ie the scope of the purported patent licence), it is a fail safe practice to assume that the licensor owns, controls and can license patents in all countries, covering:

(1) Process schemes, steps and combinations of general classes of which the particular process descriptions, methods and techniques disclosed by the licensor in his technical information supply are merely examples.

(2) Plant and equipment items of general classes, of which the specific designs and specifications supplied by the licensor for the licensee's plant are merely examples.

(3) Products of a general description of which the particular products which the licensee intends to make initially are merely examples.

(A word of explanation on point 3. The patent systems of many countries allow novel and inventive substances to be patented specifically. The patent systems of many countries also grant patent cover for known products when produced by new and non-obvious processes. Further, the patent laws of many countries will deem the importation of the direct product of a process that is patented in those countries to be infringement.)

The assumption as to extensive patent cover by the licensor has to be made by the licensee because at the time the licence is being negotiated and agreed it is normally the case that the licensee does not know details of the technology and so cannot effectively investigate the patent position himself. In any case, the licensee cannot know what may be in the patenting pipeline and as yet unpublished.

For the purpose of this analysis, it is assumed that the licensor does not have patents on any feedstocks or raw materials. It is assumed also that the licensor does not have patents covering any major end uses of products of the type the licensee will be making. But the same approach as the one adopted to cater for patents in the categories described above will be right for these areas too.

In this analysis provisions for exchange of future improvements and developments and cross-licensing under future patents that relate to these are not dealt with systematically, but, once again, the same approach can be used although there may be merit in adopting language that more closely defines rights under future patents in terms of the specific information exchanged.

The proper definition of the licensee's patent licence requires:
(a) definition of the patents and patent applications (the 'Patent Rights') under which licences of some sort (yet to be specified) are granted, and
(b) definition of the acts which the licensee is licensed to perform under those Patent Rights.

(a) DEFINITION OF PATENT RIGHTS

As mentioned above, ordinarily at the date of signature of the licence agreement the licensee does not know details of the technology and cannot know what patents the licensor controls. And so, from the licensee's standpoint, the definition should be as broad as possible. Let limitations feature in part (b), the licensed acts, so far as possible.

The definition of Patent Rights can usefully draw on other definitions; in particular Process, Products, Territory.

Assume the parties have agreed a generic process definition (the 'Process'), which is not too widely drawn in the licensor's opinion, and not too narrowly drawn in the licensee's. Assume, too, that the parties have agreed a suitable generic Products definition.

The Process might be defined as "a process for the conversion of [the raw materials types] to products comprising the steps of 1, 2, 3, etc".

A possible definition of Patent Rights, (perhaps called 'Existing Patent Rights' if there is to be future improvements exchange), might then be: 'Patent Rights' means patents and patent applications [in the Territory] [in all countries] (including divisions continuations, continuations-in-part, re-issues and extensions thereof) owned or controlled by licensor which, or the claims of which, relate to or cover the Process or Products (including process steps, methods and techniques useful therein, apparatus, plant items and equipment for use therein and [catalysts for use therein and their manufacture] [the use of catalysts but excluding catalysts per se and methods of manufacturing catalysts]), being patents or patent applications for inventions conceived prior to the date [of this Agreement] [specified in Clause...]. Patent Rights controlled by licensor include patents in respect of which licensor has the right to grant licences or immunities". The square-bracketed words are alternatives.

This formula embraces patents/applications already existing, as well as any patents which the licensor might yet decide to seek for inventions contained in his present stock of knowledge and which could, for example, cover aspects of the detailed front end engineering design he supplies to the licensee.

(National patent laws allow patenting of inventions disclosed to others in confidence, and in some cases of inventions which are being commercially worked under conditions of strict secrecy and security, and indeed, in a few countries, of inventions worked in public but abroad.)

An alternative end section to the Patent Rights definition might be "being patents or patent applications first filed prior to (date) or claiming convention priority from an application for protection in any country filed prior to (date), and any additional patents and applications in respect of the same invention but not claiming convention priority".

The date might be the agreement date or a date subsequent (eg the date of completion of the design package).

This wording may not be as satisfactory to the licensee, from the theoretical standpoint mentioned above. Faced with this wording, the licensee should, as a minimum, require the licensor to undertake not to assert any patent rights which he could have applied for but elects to defer to a later date to prevent licensee using licensor's Technical Information supplied to licensee in and for any operation of the Process [in the Plant] [in the Territory] to produce Products or against the direct or indirect supply, sale and exports of Products so made.

If the licensor desires to remain free (to the extent he has relevant patents and to the extent the law may allow) to assert patents in some countries against imports of Products made or sold by the licensee, then the Patent Rights definition will not refer to Patent Rights "in all countries" but to Patent Rights "in the Territory" or "in the following countries (named)" or "in all countries except (named)". The licensee then has no rights at all under patents in excluded areas and he is in the same position as third parties excepting, of course, as may follow from the EEC 'exhaustion of rights' principle discussed in the main text.

The above definition of Patent Rights is not directly dependent upon the definition of licensor's Technical Information to be supplied to licensee. From the licensee's standpoint this must be right because he does not want to be licensed for a production facility conforming at all times to the specific designs and instructions supplied by the licensor. He will want to modify and improve the plant and its operations, or at least to be free to do so and still enjoy freedom from suit. The Patent Rights are, of course, patents and patent applications for subject matter in the knowledge of the licensor at the time the licence is being negotiated and, so, the licensor is in a position to judge what different freedoms are being granted to the licensee by different wordings such as those given above. The words "relate to or cover" are also broad and, although commonly used, they are to some extent uncertain in scope. A narrowing of the Patent Rights definition can be achieved in a number of ways, stopping short of a mere 'label-licence' to use licensor's technical package as supplied, which can never be satisfactory for a licensee. One obvious way is to make express exceptions by patent numbers or by subject-matter categories. Another way is to use definitions based on an infringement test. Thus, Patent Rights might be defined as patents (or even claims of patents) which would be infringed by:
- the design and construction of the Plant (as defined) in the Territory in accordance with Licensor's Technical Information;
- operation of the Process (as defined) in the Plant in the Territory in accordance with Licensor's Technical Information, and/or;
- the sale in [the Territory] [any country] of Products (as defined).

This definition presumes that the Licensor does not have patents directed to the downstream detailed mechanical, electrical, instrument, and civil engineering activities for the Plant which are the responsibility of others.

Having defined qualifying Patent Rights it is now necessary to define what acts are in fact licensed under those presumed patents (or claims of patents).

(b) DEFINITION OF LICENSED ACTS

This will consist of statements such as: "Licensor hereby grants to Licensee an irrevocable, [non-exclusive] [sole] [exclusive] right and licence under Licensor's Patent Rights for the lives thereof to practise the Process [in the Plant] [in the Territory] to produce Products and the non-exclusive right and licence thereunder to supply, sell and export in and to all countries Products."

"[Licensee shall have the right to sub-license his Subsidiaries and Affiliates.]"

"Licensee [and his sub-licensed Subsidiaries and Affiliates] shall [further] have the right to sublicense customers of Licensee [and of his Subsidiaries and Affiliates] to supply, sell and export Products obtained from Licensee [or his sub-licensed Subsidiaries and Affiliates] in and to all countries."

The words in square brackets are alternatives or extras.

NOTES:
- The Territory here is the manufacturing territory (a defined term) and it would not ordinarily have the same boundaries as the Products sales area, but it could.
- The "Plant" should be defined to include modifications and extensions/enlargements, if a one-plant licence, as well as a reconstruction.
- If "sole" or "exclusive" manufacturing rights are granted for other than a short lead time (whereafter the agreement should say the rights become non-exclusive), it might be best to replace "Subsidiaries and Affiliates" by "third parties" or a qualified class of third party, or, at least, to say that sublicensing of other third parties is allowed subject to the consent of the licensor, such consent not to be unreasonably refused.
- If sole or exclusive selling rights in any named geographical markets are required (but take care!) then the statement will need appropriate modification.
- If the Products, or some of them, are patented as such by the Licensor, he may insist that trading rights apply only to Products that have in fact been made by Licensee (or sub-licensees) in Licensed Plant(s).
- The reference to "all countries" for trading rights does not override any territorial restrictions contained in the Patent Rights definition. Exports to those excluded territories are therefore subject to any enforceable patents owned there by the licensor.
- "Irrevocable" implies the licence is not a mere permit that can be withdrawn at the licensor's whim.

In addition to the immediately targeted licenced activity, what does such a patent licence allow the licensee to do without concern for any patents the licensor may have?

(1) He may modify or improve or enlarge his manufacturing plant in any way whatsoever, provided (if a one-plant licence) it is still within the Plant definition, and provided future patents (unlicensed) of licensor, or any expressly reserved existing patents of licensor, could not be enforced to prevent this.

(2) He may modify or improve the process technology provided it still stays within the Process definition, but with the same provisos as in (1) above.

(3) He may make any products if still within the Products definition (but subject to the provisos of (1)).

(4) If a production licence, he may build new plants in the Territory (subject again to the provisos of (1)) using all technical knowledge at his disposal.

(5) He may sub-license as stated.

(6) He may sell Products anywhere, subject to patents in territories excluded from the Patent Rights definition, and subject to future patents of licensor not reached by that definition.

Such actions are 'permitted' whatever the licensor's patent position may be. However, simply because the patent license does not allow the licensee to operate outside the scope of his patent licence, it does not follow that the licensor can necessarily stop the licensee doing those 'unlicensed' acts. The licensor may not in fact have any patent rights enforceable against such 'unlicensed' acts and the licensee has not contractually agreed to stay within the area of activities to which the patent licence relates. The only contractual restrictions are those discussed earlier under the section on Contractual Restrictions and Freedoms, and those restrictions relate solely to other or further use of licensor's supplied proprietary technical information.

The above 'licence' is not a warranty that all 'licensed acts' can be lawfully done. Thus, the licensee must respect third party patents, for example. In this respect it is advisable to seek from the licensor at an early stage of negotiations a statement that he has searched the patent literature and is not aware of any patent (or published application which could mature into a patent) that would prevent the construction and operation of the plant in accordance with the proposed designs and process information to be furnished by licensor, or the sale of products in [market areas]. The licensee does not, of course, need to be concerned about patents reading on to equipment and other items of plant that are purchased from industry sources of such items in the ordinary course of their business (See Chapter 4, Implied Licences).

APPENDIX III — ANNOTATED SAMPLE AGREEMENTS

In this Appendix, I set out the entire provisions of sample agreements in the main categories discussed in Chapter 4, plus an example of a technology development agreement and an example of a proprietary product evaluation agreement that might precede a new business co-operation. Alongside the agreement provisions are comments that either highlight special features or provide explanation. The agreements are modelled closely on real agreements and show, in essence, what the parties agreed after due process of negotiation. Of course, names, dates, technical subject matter and other incidental details have been changed, or simply blanked out. The agreements are, I believe, reasonably representative of their class and the origin, context and content of their real antecedents were known to me personally. The agreements are not offered as templates and their provisions are not to be regarded as standard clauses. I refer the reader back to the caution in the opening paragraph of Chapter 5.

Section 1 is a bare patent licence in an EEC context.

Section 2 is a patent licence with ancillary technical information exchange.

Section 3 is a horizontal cross-license of production experience between two manufacturers of the same basic product by similar technologies.

Section 4 is a joint development agreement between parties whose relevant knowledge and expertise are complementary as in a vertical joint development of a commercial application of a new class of product by a producer of the product and a prospective converter of such product into fabricated or compounded items.

Section 5 is a evaluation agreement between a producer of a proprietary development material and a downstream user or consumer of similarly functioning materials who desires to evaluate the new material for his applications.

Section 6 is a total technology licence for a commodity product comprising a licence agreement, a plant design agreement, a project technical assistance agreement and an operator training agreement.

SECTION 1 — BARE PATENT LICENCE

PATENT LICENCE AGREEMENT

This Agreement is made as of the lst day of January, 1990 by and between Patent Holdings PLC, a company registered in England with its registered office at .... (herein referred to as 'PH') and Electric Appliances Limited, a company registered in England with its registered office at .... (herein referred to as 'EAL') and German Control Systems AG, a company registered in Germany, with its principal business address at .... (herein referred to as 'GCS').

Whereas

A. EAL is a wholly owned subsidiary of PH and owns an electric appliances business.

B. PH is the registered proprietor of a group of patents (and applicant of one pending patent application) in certain European Countries relating to electric control systems and their use as controllers of electric appliances (herein called 'the Patent Rights' and defined in Article 1.1 hereof).

C. GCS has requested a licence under the Patent Rights.

NOW, THEREFORE, the parties hereto agree as follows:

SECTION 1: COMMENTS

Preamble: A tripartite agreement. PH, the parent of EAL (see Recital A), owned the relevant patent rights but EAL owned and ran the business in control systems and appliances. EAL and GCS knew each other; they would operate the agreement.

Recital B: Patent cover in Europe was not comprehensive (selected filings only). The pending application was involved in protracted Patent Office examination.

ARTICLE 1: DEFINITIONS

1.1 'Patent Rights' means the issued patents and the pending patent application listed in Schedule 1 hereto (and any patent as may be granted on the pending application).

1.2 'Territory' means all those European countries, territories and areas within which the Patent Rights, collectively, have effect.

1.3   'EEC' means member countries of the European Economic Community and includes countries which after 1 January 1990 become full members of the European Economic Community as from the date of their accession.

1.4   'Electric Control System' means an electric control system possessing an internal electronic output sequence controller that in response to differentiated input control signals effects preassigned patterns of operation of multi-channel output switch gear according to the principles broadly described in the specifications of the Patent Rights and on which the claims thereof rely.

1.5   'The date hereof' means the 1st January, 1990.

1.6   'Make' and 'manufacture' and similar expressions include assembly from GCS-specified components acquired from others.

1.7   'Net Sales Value' and 'NSV' shall mean the gross invoice value of Electric Control Systems sold by GCS or a sublicensed Subsidiary less the cost of discounts (other than prompt payment discounts), transportation, insurance, duties, taxes assessed directly on sales and allowances to customers except that for sales to a GCS Subsidiary for its use the Net Sales Value shall be deemed to be what it would have been had the sale been made to an independent third party in the country of the said Subsidiary as reasonably indicated by consideration of comparable actual third party sales.

1.8   'Subsidiary' means any company of which PH or GCS, as the case may be, owns directly or indirectly more than 50% of the voting stock of the company.

1.9   'PH Affiliate' means a company of which PH owns directly or indirectly from 35% to 50% of the voting stock of the company.

ARTICLE 1: COMMENTS

1.1   Note that the UK patent was included. It was judged that the 'exhaustion of rights' doctrine within the EEC made it pointless to try to isolate the UK because goods put on the market in Continental Europe cannot be excluded from the UK. Anyway, the whole European market for the novel control system needed development. GCS would help in that.

1.2   The Territory is the patented area of Europe. There were non-European patents; they were not being licensed. Further there was no know-how licence, so there was no need to cater for non-patented countries of Europe.

1.3   This is relevant to Article 4.

1.4   This is probably gibberish! But the final bit reflects the fact that the patent claims differed significantly in the different countries. This is umbrella language but not so wide as to cause problems with Article 4.

1.6   See Article 2.

1.7   The final proviso is needed to deal with peculiar intra-Group transfer prices. If there is a significant external reference market, and a basis of trust, the language will work.

1.8   Controllable companies.

1.9   Relevant only to Article 4.

ARTICLE 2: THE LICENCE GRANTS

2.1   PH, at the request of EAL, hereby grants to GCS the non exclusive right and licence under the Patent Rights to manufacture in the EEC Electric Control Systems and to use them or supply them to others for use in electric appliances in the Territory.

2.2   GCS may sublicense its Subsidiaries. GCS and sublicensed Subsidiaries may have Electric Control Systems made for them by others and may sublicense direct, intermediate, and ultimate customers to distribute for use and to use in the Territory Electric Control Systems made by, or for, GCS and sublicensed Subsidiaries.

Neither PH nor EAL shall assert the Patent Rights, or any of them, against direct customers of GCS and its Sublicensed Subsidiaries for Electric Control Systems where the transactions themselves give rise to a liability under Article 3.2 or where GCS can show that the transactions in question have been included in a return made, or will be included in a return to be next made, under Article 3.3.

EAL agrees with GCS that it will consult with GCS before EAL or PH asserts any of the Patent Rights against a possessor or user of Electric Control Systems that originated from GCS or a GCS Subsidiary. No other right, licence, power to sublicense, or immunity is granted or implied in respect of the Patent Rights, patents outside the Territory corresponding to the Patent Rights, or other patents of PH or EAL.

2.3   For the avoidance of doubt, this Licence Agreement does not imply any commitment or agreement by GCS not to make or sell or use Electric Control Systems in any country where no patent corresponding to the Patent Rights exists or where the relevant Patent Rights have lapsed, been surrendered or revoked or have expired.

2.4   GCS shall continue to enjoy the rights and licences granted in Articles 2.1 and 2.2 during the term of this Licence Agreement. The undertaking for the benefit of customers and users in Article 2.2 shall be irrevocable and shall survive for the lives of relevant Patent Rights.

2.5   In consideration for the rights and licences granted in this Article 2, GCS shall make payments in the amounts, at the times and by the routing specified in Article 3.

2.6   GCS will use its reasonable endeavours to exploit the rights and licences hereby granted through manufacture and sales of Electric Control Systems as the market therefor admits.

ARTICLE 2: COMMENTS

2.1   This does not prevent sales outside the Territory but where there are relevant patents PH/EAL can assert them against GCS (see last paragraph of 2.2). See Article 4 also.

2.2   The sublicence through the distribution chain to the ultimate customer was needed because of (i) partial geographical patent cover in Europe and (ii) some patents of the patent family only claimed specific appliances incorporating the novel controller, or specific methods of controlling electric appliances.

2.3   There are no purely contractual manufacturing or sales restrictions.

2.4   Once sold, always licensed. See Articles 5.2 and 5.3.

2.6   Obviously not a sword against GCS but of some merit if GCS should lose interest in promoting the new control system. The licence is non-exclusive and Article 4 has relevance, so PH/EAL are not too exposed.

ARTICLE 3: PAYMENTS

3.1   The sum of DM 100,000 shall be paid by GCS to EAL within 30 days following the date of receipt by GCS of its fully executed engrossment of this Licence Agreement and a further sum of DM 100,000 shall be paid by GCS on each subsequent anniversary of the date hereof during the term of this Licence Agreement until at the latest the tenth anniversary. Payments shall be made by the routing specified in Article 3.6.

3.2   The sum in DM equal to 5% of NSV shall be due as a royalty from GCS to EAL in respect of Electric Control Systems made and used or supplied for use with the benefit of the licences, sublicences and immunities herein granted. Credit shall be allowed to GCS against the aggregate royalty sum due and payable in respect of any calendar year during the term of this Licence Agreement of the sum paid in that year pursuant to Article 3.1.

3.3   Payment of a royalty sum due under Article 3.2 (after allowing for the said credit) shall be made within 30 days following the end of the calendar year in which the liability for that royalty sum was incurred or following termination in the case of the calendar year in which termination of this Licence Agreement occurs. For the avoidance of doubt, only one royalty is payable in respect of any one Electric Control System. Thus, for example, royalty on Electric Control Systems whose manufacture was (sub)licensed would fall due upon their use or earlier first sale, as the case may be, wherever that use or sale takes place. Subsequent (sub)licensed distribution into or within the Territory would not incur any liability for a further royalty. In the case of Electric Control Systems made without the benefit of this licence (for example, in Holland where no patent corresponding to Patent Rights exists) and which thereafter are distributed for (sub)licensed use or sale within the Territory, the royalty would fall due when they were so used or first sold, as the case may be.

3.4   Stocks of Electric Control Systems made under the licence herein granted (or under a sublicence) and which are held by GCS and its Subsidiaries at termination of this Licence Agreement shall, if liability for royalty thereon has not previously arisen under Article 3.2, be royalty bearing as if they had been used under licence by GCS immediately prior to termination, and such royalty shall be paid within 30 days following termination.

3.5   GCS shall keep (and require sublicensed Subsidiaries to keep) full and accurate records of all usages, sales and other supplies of Electric Control Systems as necessary for determination of sums due under Articles 3.2, 3.4 and 3.9.

The records shall identify the parties and places to which sales and supplies are made, the numbers of Electric Control Systems sold or supplied, and the dates of all transactions. Such records shall be made available for inspection during normal business hours and once in each calendar year by an independent auditor selected by EAL but subject to approval by GCS, which shall not be unreasonably withheld, to verify the accuracy of the payments made pursuant to Articles 3.2, 3.4 and 3.9 during the three previous calendar years. The auditor shall disclose to EAL only that the royalties which should have been paid have been paid, or, if there is a discrepancy, report to EAL the amount of any under- or over-payment. Any under-payment shall be immediately paid by GCS and any over-payment shall be credited against the next royalty payment.

3.6   Payments of sums and royalties payable hereunder shall be made to account No. 1234567 of PH Finance PLC at Deutsche Bank AG, Frankfurt. GCS will within 14 days of each payment being made notify EAL of the amount.

3.7   Sales effected in a currency other than DM shall be converted into DM by using the rate of exchange at which the bank specified in Article 3.6 will purchase that currency for DM at the close of business on the last business day of the calendar month in which the sales were effected.

3.8   All taxes, levies and other similar payments payable outside the United Kingdom on sums due and payable under this Agreement shall be borne and paid by GCS except to the extent that credit may be obtained for such taxes, levies or other payments against United Kingdom tax payable on such sums either (i) under a double taxation convention between the Government of the United Kingdom and the Government of the Federal Republic of Germany, or (ii) under United Kingdom legislation for granting credit unilaterally.

The sums payable hereunder are exclusive of UK Value Added Tax (where applicable) and GCS undertakes to bear and pay any UK Value Added Tax which may be chargeable.

3.9   Electric Control Systems made and consumed internally within GCS or a GCS Subsidiary (so that there is no external supply or sale) shall attract royalties. They shall be assigned an NSV for royalty calculation purposes, and royalty shall be due and payable in respect thereof, as if they had been supplied at the time of their use to a GCS Subsidiary based in the country of use.

3.10   If GCS or a GCS Subsidiary shall for the purposes of bona fide market development or customer evaluation supply Electric Control Systems at less than the full production cost (inclusive of packaging) to GCS (or its Subsidiary sublicensed for manufacture) of those Electronic Delay Detonators, they shall be deemed, for the calculation of payable royalties, to have been supplied at an NSV equal to that full cost.

ARTICLE 3: COMMENTS

3.1   Minimum annual payments but creditable against royalties (see 3.2).

3.2/3.3 Royalty payable on control systems used or sold under licence, not when manufactured (even if made under licence). Only one royalty per item.

3.4   This was needed to pick up stocks because of the way 3.2/3.3 were worded. See 3.9.

3.5   Note that the auditor (but not EAL) has access to GCS commercial information. See also Article 5.1.

3.6/3.7 Necessary detail.

3.8   Some such clause as this is vital to protect the licensor, but different companies' tax experts have their own preferred language. This is a 'grossing-up' formula reducing the overall tax impact to what the position would have been had GCS been a UK Company. For a licence to distant lands the VAT provision would be a little excessive.

3.9   See 3.4.

3.10   This clause recognises that GCS may use price incentives to develop the new business. In view of the minimum annual payment, this clause was not expected to have much impact.

ARTICLE 4: THIRD PARTIES

4.1   In this Article the expression 'third party' does not include any PH Subsidiary or PH Affiliate.

4.2   PH undertakes, in recognition of the effort and resources that will be required to be expended by GCS and/or its Subsidiaries to develop the European Market for Electric Control Systems, that PH will not voluntarily license any third party under the Patent Rights to supply into the EEC Electric Control Systems made outside the EEC. This undertaking will be binding for an initial 3 years from the date hereof and will continue to apply from year to year thereafter unless cancelled by six months' notice from GCS to EAL.

ARTICLE 4: COMMENTS

4.1/4.2   This Article was a needed concession to GCS who were content to face competition from within the EEC and from the PH Group of Companies but did not wish to develop the EEC market only to find that PH/EAL then licensed an outsider to sell in to the EEC.

ARTICLE 5: GENERAL

5.1   This Agreement shall continue in force until the last remaining patent of the Patent Rights has expired, been revoked, surrendered or allowed to lapse unless earlier terminated by notice under the provisions of this Article 5. The provisions of Article 3.5 shall however continue in force beyond any termination for a further 2 years.

5.2   GCS shall be entitled to terminate this Agreement at any time by 2 years notice in advance.

5.3   If GCS shall be in breach of any of its obligations herein or, if the breach is capable of being remedied, shall fail to remedy the breach with all practicable speed upon receiving written notice of the breach (and in any event within 60 days of such notice) PH or EAL shall have the right to terminate this agreement by giving six months' notice of termination.

5.4   Neither PH nor EAL warrants the validity of the Patent Rights or any of them. For the purpose of this Agreement any amendment, curtailment, revocation, lapse, or abandonment of the Patent Rights, or any of them, shall not have retrospective effect. In no event shall sums and royalties paid to EAL be refunded to GCS and those due to EAL shall be paid notwithstanding that, had any occurrence as stipulated above in this Article 5.4 in relation to the Patent Rights, or any of them, taken place at some earlier time, such payments would not then have fallen due.

5.5   This Agreement may not be assigned by any party without the written consent of the others. Consent shall not be unreasonably refused to any proposed assignment to a wholly-owned Subsidiary or, in the case of EAL, to PH or any other wholly-owned Subsidiary of PH. Provided that in the case of an assignment by PH ownership of the Patent Rights shall also be assigned to the assignee.

5.6   Notices shall be sent by registered mail or recorded delivery and addressed to the EAL Company Secretary, in the case of notices from GCS to EAL or PH, and to the Patents Department Manager, in the case of notices to GCS, at the corporate addresses above written. Notices to EAL under Article 3.6 may, however, be sent by ordinary first class airmail, fax or telex.

5.7   All disputes or differences which may arise out of, in relation to or in connection with this Agreement shall be settled amicably between the parties hereto by discussion between them and with all practicable speed.

However, in case any dispute or difference is not settled, each party shall have the right to refer it to arbitration for final settlement in conformity with the rules of conciliation and arbitration of the International Chamber of Commerce, Paris, by three arbitrators appointed in accordance with such rules being in force at such time.

The arbitration shall take place in Zurich/Switzerland. The procedural laws of the place of arbitration shall be applied. English shall be the language used in proceedings.

The arbitration award shall be binding on the parties who shall act accordingly. In the course of arbitration the parties shall continue to execute their obligations under this Agreement except for those directly the subject of the Arbitration.

The costs of arbitration shall be borne as the arbitrators shall specify in the arbitration award. The parties will, however, bear their own internal legal and administrative costs.

5.8 This Agreement shall be governed by the English text and shall be subject to and interpreted in accordance with the substantive law applicable to resident companies of the Canton of Zurich (Switzerland) except that questions concerning the status, scope or effect of a patent shall be decided according to the law of the country issuing the patent. Any amendment or modification of this Agreement shall be in writing and signed by duly authorised representatives of the parties.

AGREED: AGREED:
By: .... By: ....
Date: .... Date: ....

AGREED:
By: ....
Date: ....

ARTICLE 5: COMMENTS

5.1 – 5.3 Nothing extraordinary here

5.4 This is a contractually important clause. The validity of patents is always uncertain until fully tested. GCS were free of course to assess the patents' validity before agreeing to take a licence. They can see the scope of patents issued and are not stopped from challenging the patents or raising with PH/EAL any matter affecting validity that may yet come to light. They can terminate the licence, albeit on 2 years' notice. What this clause recognises is that while the patents exist they have deterrent value opposite unlicensed competitors and the licence is a valuable freedom from suit, at its lowest.

5.5 Sensible freedoms and safeguards.

5.7/5.8 ICC Zurich disputes settlement. A respected body, neutral territory. But what is there in this Agreement that will get that far? Note the neutral choice of law and the requirement that patents be assessed according to the laws of the issuing states.

SCHEDULE 1: PATENT RIGHTS

Country & Number   Expiry Date
UK
IT
DT
ES
FR
OE (Austria)
SW
                (application)
                    pending

SECTION 2: PATENT LICENCE AGREEMENT

This agreement is made the lst day of January, 1990 by and between the Able-Bodied Company of .... , UK, (herein referred to as 'AB') and The Can-Do Company of .... , UK (herein referred to as 'CD').

Whereas AB has developed and is seeking patent protection for a water purification system details of which have been disclosed to CD under a Secrecy Agreement dated lst June 1989, and whereas CD has now requested a licence to manufacture, operate and sell such system.

NOW, THEREFORE, the parties hereto agree as follows:

SECTION 2: COMMENTS

AB had developed to a practically operable state the subject water purification system for its own internal purposes, being a large consumer and user of water, but was not in the business of designing and supplying particular systems suitable for use by others. CD was.

The pre-licence evaluation convinced both parties that there was a market for the systems and that the prospects for worthwhile patents were good.

ARTICLE 1: DEFINITIONS

1.1   'Patent Rights' means the patent applications listed in Schedule 1 hereto and patents as may be granted thereon, including any renewals, reissues, continuations or continuations-in-part.

1.2   'Territory' means all those countries, territories and areas within which the Patent Rights, collectively, have effect. 'UK' means the United Kingdom.

1.3   'System' means a water purification system comprising the following units ....

1.4   'Station' means an installed water treatment plant incorporating the System.

1.5   'Technical Information' means all technical information, know-how, and experience developed or acquired by a party hereto prior to the 5th anniversary of the date hereof and at that party's free disposal which relates to the design, installation, performance, repair or maintenance of Systems.

1.6   'The date hereof' means the date first written above.

ARTICLE 1: COMMENTS

1.1   At the date of this Agreement, details of the patent applications in the European Office had not yet been published under the early publication procedures.
        Continuations-in-part are a feature of US patenting where the applicant builds on his early priority application with additional technical disclosures. These are at the election of AB, of course, who might instead, in the case of a significant development, proceed with an entirely independent patent application. C-I-P's are included here because there is ongoing information exchange for 5 years (see 1.5 and Article 3).
        Schedule 1 listed all relevant patent applications on file at the Agreement date.

1.3   The definition mirrored the main (broadest) claim of the European Patent Application. Curtailment of the scope of the patent claims during prosecution would not matter to CD because of the language chosen in the Licence Grants (Article 2) and the Payments Article (Article 4).

1.4   The patent application also included claims for water-treatment plant incorporating the System.

1.5   This embraces all AB knowledge at the Agreement date plus experience either party obtains concerning the system during the next 5 years.

ARTICLE 2: THE LICENCE GRANTS

2.1   AB grants to CD the non-exclusive right and licence under the Patent Rights to manufacture in the UK Systems and components thereof and to supply the same directly or through its agents to others in the Territory or elsewhere where such manufacture or supply would, but for such right and licence, constitute direct or contributory infringement of any Patent Rights. 'Manufacture' includes manufacture for and on behalf of CD.

2.2   AB grants to CD the right and licence to install and operate on Stations in the Territory Systems manufactured and supplied by CD where the installation and/or operation would, but for such right and licence, constitute infringement of any Patent Rights.

2.3 AB grants to CD the right to grant to Station owners or operators who acquire Systems manufactured and supplied by CD the right to install and operate such Systems, where the installation or operation would, but for such right, constitute infringement of any Patent Rights.

2.4 CD shall continue to enjoy the rights and licences granted in Articles 2.1 and 2.2 during the life of this Agreement. Rights granted by CD to Station owners and operators pursuant to Article 2.3 shall be irrevocable and shall survive for the lives of relevant Patent Rights.

2.5 In consideration for the rights and licences granted in this Article 2, CD shall make payments to AB in the amounts and at the times specified in Article 4.

2.6 Insofar as the law of any part of the Territory shall not recognise the grant of rights and licences under a patent application, the aforesaid grants to CD shall, during the pendency of any such patent application, take effect as an undertaking by AB not to assert any right arising on the grant of a patent on such application against any manufacture, supply, installation or operation as referred to above in this Article 2 occurring prior to patent grant, and as a commitment by AB to grant to CD the rights and licences aforesaid under patents resulting as of the date of grant thereof. Article 4 (Payments) shall take effect as if the fees due thereunder were in consideration for such undertaking and commitment by AB in respect of any part of the Territory as aforesaid during the pendency of the patent application therein.

ARTICLE 2: COMMENTS

2.1 – 2.3 This is language appropriate for issued patents. Published but still pending patent applications put the world on notice that patents may issue but until granted as patents they are not enforceable and their eventual scope is uncertain. 2.6 addressed this interim position.

The licence concerns UK Systems manufacture and global sales. The licences allow what otherwise would infringe. Of course that is all that can be licensed but this language is not redundant because it is the test of whether payments are due under Article 4.

The different grants of 2.1, 2.2 and 2.3 reflect the claim structure of the patent applications, the fact that a System would be made in the UK and shipped for installation and use in another and the fact that CD had ideas of providing a water-treatment service.

2.3 grants CD a right to license particular installations.

2.4 Installed Systems, and the Stations of which they are part, are always licensed.

2.6 As between AB and CD this is clear enough. It establishes a legal basis for payments to AB pending patent grant and provides necessary comfort to CD and (practically, if not legally) to CD's customers.

ARTICLE 3: EXCHANGE OF TECHNICAL INFORMATION

3.1 AB and CD shall exchange their respective Technical Information. The exchange shall be effected in a timely and mutually convenient manner and, as a minimum, representatives of the parties shall meet one day each year of this Agreement, up to and including the 5th year, to exchange and discuss their respective Technical Information.

3.2 Each recipient of the other's Technical Information shall be entitled to make free and unrestricted use of such Technical Information for all purposes, provided that, if such Technical Information, or any use thereof, is or shall become, after the date hereof, the subject of any intellectual property rights of that other party in any country (additional to the Patent Rights), the recipient party shall have an irrevocable royalty-free right and licence, with sublicensing rights, under such intellectual property rights limited to the use of such Technical Information in and for the manufacture, use, supply and supply for use of Systems.

3.3 Each recipient of the other's Technical Information shall treat that Technical Information, as regards disclosures to others, in the same manner and according to the same policies for the due protection of secret, confidential or proprietary information as it treats its own Technical Information of similar status and subject matter.

ARTICLE 3: COMMENTS

3.1   The parties had a mutual interest in this exchange. A lightweight clause was judged good enough.

3.2/3.3   Exchanged information becomes part of each party's stock of knowledge, to be dealt with responsibly but otherwise without restriction. The patent application when published disclosed much of AB's relevant pre-existing information that had previously been covered by the June '89 Secrecy Agreement. On this formula, if AB or CD should devise an improvement that is to be patented, a patent application should be filed (or a specific secrecy undertaking extracted) before it is exchanged. Note the relevance of the definition of 'System' to rights under any improvement patents (or other intellectual property rights) obtained.

ARTICLE 4: PAYMENTS

4.1   In respect of manufacture and supply pursuant to the right and licence granted in Article 2.1, a fee of £X000 (adjusted as specified in Article 4.5) shall be due from CD to AB in respect of every System so manufactured and supplied (either factory-assembled or as separate components for assembly elsewhere).

4.2   In respect of installation and/or operation by CD at Stations of Systems pursuant to the right and licence granted in Article 2.2, a fee of £X000 (adjusted as specified in Article 4.5) shall be due from CD to AB in respect of each Station for every System so installed and/or operated PROVIDED THAT, if a fee shall already be due to ICI under Article 4.1 in respect of any System so installed and/or operated, that fee shall be credited against fees due under this Article 4.2.

4.3   In respect of the grant to Station owners or operators of the right to install and operate Systems pursuant to Article 2.3, a fee of £X000 (adjusted as specified in Article 4.5) shall be due to AB from CD in respect of each System so installed and/or operated PROVIDED THAT, if fees shall already be due to AB under Article 4.1 or 4.2 in respect of that Station and that System, no fee under this Article 4.3 shall be due to AB.

4.4   Fees due under Article 4.1, 4.2 or 4.3 shall be payable within 90 days following the date on which the System is delivered (or delivery of the components of it is complete) or on which the System is invoiced to the owner/user, whichever is the earlier date.

4.5   On July 1, 1990, and subsequently on the first day of July in each following calendar year the sum of £X000 named in Articles 4.1, 4.2 and 4.3 shall be automatically adjusted according to the formula:

Adjusted Sum = £X000 x IR/IO

where: I is the Index of Average Earnings, All Industries (Whole Economy) published in the Department of Employment Gazette by HMSO, IO is the value of I for September, 1989 and IR is the value of I for March which immediately precedes the July adjustment date.

4.6   CD shall keep full and accurate records of all sales and supplies of Systems, and components thereof, identifying the parties and places to which sales and supplies are made, the items sold or supplied, and the dates of all transactions. Such records shall be made available for inspection during normal business hours by an independent auditor appointed by AB solely so that he may either confirm to AB that fees which should have been paid have in fact been paid or, if there is a discrepancy, report to AB the amount of any under- or over-payment. Irrespective of any such independent audit, CD shall before the end of February in each calendar year during the life of this Agreement report to AB the names of parties, locations and dates of all transactions giving rise to fees payable to AB in the previous calendar year ending December 3lst.

4.7   Value Added Tax on fees payable to AB hereunder shall be payable by CD as required by law and shall be a strictly nett extra charge. So far as possible, CD will adopt permitted self-billing arrangements and will pay such taxes with the fees to which they relate.

ARTICLE 4: COMMENTS

4.1   This concerns Systems sold to others. Pending issuance of the patents, Article 2.6 governs this clause and clauses 4.2 and 4.3. If what is made and supplied meets the definition of 'System', fees are due (Payment is dealt with later). Following patent grant, fees are due if manufacture and supply of the System required the patent licence in Article 2.1. The parties were content to apply these rules and the agreement provides checks and balances in Article 5.

Note that the parties could not know what scopes of patents would be eventually allowed in the different countries.

4.2   This concerns installation by CD or operation by CD at Stations, possibly at CD's plant or as a service to others. If installation or operation requires the patent licence of Article 2.2, a fee per System per Station is due, with a credit for fees payable in respect of those Systems under 4.1.

4.3   This picks up the situation where the only activity in the chain of manufacture, supply and use of a System that needs a patent licence is operation at a Station.

4.4   This specifies when payments are to be made. Both parties are UK companies and fees are sterling sums not related to any other currency.

4.5   An inflation hedge.

4.6   Note annual reporting of transactions (but not price information).

4.7   In a UK transaction, VAT has to be dealt with.

ARTICLE 5: GENERAL

5.1   This Agreement is an agreement made in England and subject to English Law, save that any question or dispute concerning the validity of any of the Patent Rights (or any claims thereof) or as to whether any act would, apart from the grants herein, constitute direct or contributory infringement of any of the Patent Rights shall be determined according to the patent laws, codes, and statutes of the jurisdiction in which those Patent Rights have effect.

5.2   This Agreement shall continue in force from the date first-above written until the expiry of the Patent Rights or until such earlier date as this agreement is terminated by notice of either party under the provisions of this Article 5.

5.3   CD shall be entitled to terminate this Agreement at any time by notice in writing. All fees due to AB at the date of termination under any provision of this Article 5, shall be paid forthwith.

5.4   If CD shall be in breach of any of its obligations herein or, if the breach is capable of being remedied, shall fail to remedy the breach with all practicable speed upon receiving written notice of the breach (and in any event within 60 days of such notice) AB shall have the right to terminate this agreement by giving six months' notice of termination in writing.

5.5   If no fees under Article 4 shall have fallen due to AB in the period up to, at the earliest, the second anniversary of the date hereof, AB shall have the right to terminate this agreement forthwith by notice in writing PROVIDED THAT, if CD shall give notice to AB with documentary corroboration that it had at the date of receipt of the aforesaid notice become liable for payment of fees to AB or had entered into binding contractual supply commitments which will give rise to fees being payable to AB or had made and was continuing to make substantial efforts to promote and sell the Systems, such notice by AB shall be deemed not to have been given.

If, conversely, payment of fees to AB under Article 4 shall cease for a continuous period of 12 months at any time after the second anniversary of the date hereof, AB shall have the right to terminate this Agreement by notice in writing, such termination to take effect on the second anniversary of the date of giving the notice.

5.6   AB does not warrant the validity of the Patent Rights or any of them. For the purposes of this Agreement any amendment, curtailment, revocation, lapse, or abandonment of the Patent Rights, or any of them, shall not have retrospective effect. In no event shall fees paid to AB be refunded to CD and fees due to AB shall be paid notwithstanding that, had any occurrence as stipulated above in this Article 5.6 in relation to the Patent Rights, or any of them, taken place at some earlier time, such fees would not then have fallen due.

5.7   Without prejudice to CD's right in Article 5.3, if any question or dispute of the kind referred to in Article 5.1 shall arise which the parties are not able to resolve quickly to their mutual satisfaction by discussion between them, the question or dispute shall be submitted to an independent patent practitioner (being an agent, attorney, counsel or similarly qualified person) knowledgeable and experienced in the patent law and practice most relevant to the matters in issue for his opinion. CD shall select the person to give this opinion from a list of at least four independent nominees provided by AB. The costs of this referral shall be shared equally by the parties. The opinion given shall not be binding on the parties, but shall serve as the basis for a further bona-fide attempt by the parties to resolve matters by negotiations to their mutual satisfaction.

5.8   This Agreement is personal to the parties and shall not be assigned by either party without the written consent of the other. Consent shall not be unreasonably refused to any proposed assignment to a wholly-owned subsidiary or, in the case of CD, to its parent company or any other wholly-owned subsidiary of such parent company. Further, AB shall not unreasonably refuse consent to CD sublicensing a wholly-owned subsidiary of CD or of its parent in respect of any of the rights and licences herein granted.

5.9   The parties will at the request of either from time to time review the working of this agreement (including the levels of fees prescribed) to consider whether or not any changes or adjustments should desirably be made in the light of market experience.

5.10   Notices shall be sent by registered mail or recorded delivery to the addresses for the parties first written above and addressed to the Secretary.

AGREED: AGREED:
By: ....   By: ....
Date: ....   Date: ....

ARTICLE 5: COMMENTS

5.3   CD can immediately terminate by notice but AB then can assert the patents against future activity by CD.

5.4   In practice this deals with non-reporting and non-payment.

5.5   The granting of this licence made it more difficult to interest others in taking a licence and if another had been licensed CD's incentive to develop a market would have been reduced. The first paragraph begs the question what constitutes 'substantial efforts' but AB still felt the clause had merit.

5.6   This clause was discussed in Section 1 of Appendix III.

5.7   Self-explanatory. The final sentence is peculiar but made sense for this relationship.

5.8   It is always important to deal explicitly with rights to assign and Sublicense.

5.9   This is cosmetic. It reflected the lack of sound data at the date of the Agreement of the sort of price the System might command in the market and uncertainty over whether foreign business could sensibly for long be supplied from a UK manufacturing base.

SCHEDULE 1
EPA No ..... designating
UK
Belgium
France
Germany
Netherlands
Italy
and equivalents filed in Australia, Canada, Norway and USA.

SECTION 3: INFORMATION CROSS-LICENCE

THIS AGREEMENT effective the 1st day of January One thousand nine hundred and ninety BETWEEN AB COMPANY an English Company, whose registered office is situated at .... London, England (hereinafter called 'AB') of the one part and CD Inc a Corporation of the State of New York, USA having a place of business at .... (hereinafter called 'CD') of the other part.

WHEREAS:

1. Both parties are engaged in research and development relating to, and the commercial manufacture of, lubricant additives and they each possess secret and confidential technical information, data and knowhow on their structure, performance properties and their production on a commercial scale.

2. The parties have agreed to effect an exchange of such information, data, and knowhow and to grant rights to each other for the use thereof on the terms and conditions hereinafter set out.

NOW THEREFORE IT IS HEREBY AGREED as follows:

In this Agreement the following terms shall bear the meanings hereby assigned to them:

COMMENTS

Recitals: This was a horizontal relationship — AB and CD were in similar businesses in their geographically distinct markets. They were not direct competitors. Their products were performance products, ie required to do a particular job. Their product portfolios were similar, as were their production processes. They expected that by sharing information correlating product structure, performance and production technology the quality and efficiency of their operations would be improved and they would have a broader base of data and experience with which to confront the challenges of the market place.

ARTICLE 1: DEFINITIONS

1.1 'Subsidiary' shall mean a commercial entity in which a party hereto owns directly or indirectly more than fifty percent (50%) of the equity interest and, in the case of AB, additionally includes Lubricant Blenders Ltd.

'Named Subsidiaries' means, in the case of AB, ... and, in the case of CD, ...

1.2 'Joint Venture' means production plant for the commercial manufacture of lubricant additives owned, jointly with others, by a party hereto or its Subsidiary and operated on behalf of the joint owners by the party or its Subsidiary.

1.3 'Products' means the lubricant additives identified by their current trade designations in Schedule 1 hereto, and such further products as the parties may from time to time agree to include in Schedule 1.

1.4 'Technical Information' means all technical information, data and knowhow concerning and relating to the structure, performance properties, and commercial production of Products as further specified in, and with the exclusions recited in, Schedule 2 hereto.

1.5 'Cut-off Date' means the date specified by either party to the other by notice in writing for the purposes recited in Article 2 hereof, or, if no such date is specified, the date on which this Agreement is terminated.

ARTICLE 2: COMMENTS

1.1 Subsidiaries may be sublicensed (See Article 4). Named Subsidiaries' patents are brought within the patent non-assertion undertakings (see Article 5).

1.2 A Joint Venture is a jointly owned plant, not a legal entity. Increasingly there is a need to contemplate this sort of co-operative activity.

1.3 This defines the products for which there will be an exchange of information (see 1.4 and Article 2.1). Note that additions to the list are contemplated.

1.4 Schedule 2 was an important document because it defined the precise scope, depth, and quality of the intended exchange of information.

1.5 See Article 2.1. This was a key safeguard in case there should be a perceived imbalance in the quality and relevance of the initial exchange or if the parties' production technologies and products should later fundamentally diverge.

ARTICLE 2: INFORMATION EXCHANGE

2.1   AB and CD will during the currency of this Agreement each make available to the other all Technical Information in its possession at any time prior to the Cut off Date which the disclosing party is legally free to disclose to the other for use in respect of Products. If either party shall elect to specify the Cut-Off Date by notice it shall not be a date sooner than the second anniversary of the date hereof nor less than 6 months after the giving of notice. Technical Information shall not include market and marketing information nor information disclosing either party's capacity for production or actual production of Products.

2.2   Each of AB and CD agrees that if it makes any agreement with any third party at any time during the currency of this Agreement which results in it being constrained in its freedom to make Technical Information available to the other party to this Agreement pursuant to the terms hereof, it will notify the other party giving all relevant details and said other party shall have the right to terminate this Agreement by notice with immediate effect.

ARTICLE 2: COMMENTS

2.1   Note: information existing prior to the Cut-Off Date; only that which can be disclosed for use; 'commercial' information excluded.

2.2   See Article 9. An attempt to deal with a fundamental change in surrounding circumstances.

ARTICLE 3: RIGHTS TO USE EXCHANGED INFORMATION

3.1   Subject to the provisions of Article 5.1 below, AB grants to CD a non-exclusive right to use Technical Information received hereunder in the research, development, manufacture (in plant owned by CD or in its Joint Venture) and sale of Products and other lubricant additives.

3.2   Subject to the provisions of Article 5.2 below, CD grants to AB a non-exclusive right to use Technical Information received hereunder in the research, development, manufacture (in plant owned by AB or in its Joint Venture) and sale of Products and other lubricant additives.

ARTICLE 3: COMMENTS

3.1/3.2   Having defined carefully the information to be exchanged, the 'cross-licences' extend to all activities in the business area. A more restricted 'licence' would have been unmanageable and unpoliceable.

ARTICLE 4: DISCLOSURE AND LICENSING INFORMATION

Each party shall have the right to disclose Technical Information received hereunder to, and to grant non-exclusive rights to use the same in the research, development, manufacture (in own plant or Joint Venture) and sale of lubricant additives to (i) its Subsidiaries and (ii) Undertakings to which it is disclosing, and granting user rights in respect of, its own Technical Information for the purposes of the production or use of lubricant additives, insofar as the Technical Information received hereunder has become inextricably intermingled with such own Technical Information and does not form the major portion of the whole, or is information which the originating party makes freely available to its customers in aid of use and sales of lubricant additives.

ARTICLE 4: COMMENTS

This clause reflects the business need. The 'Undertakings' referred to in (ii) were companies with which either party had a licensing agreement. This provision was enabling; the real safeguard was the fact that neither party had an interest in uncontrolled dissemination of information.

ARTICLE 5: PATENT NON-ASSERTION

5.1   AB undertakes, on behalf of itself and its Named Subsidiaries, not to assert against CD present and future patents owned by AB and its Named Subsidiaries in all countries insofar as such patents are for inventions contained in, or exemplified by, Technical Information disclosed to CD hereunder, to prevent the development, manufacture and sale of lubricant additives. Additionally, AB will use reasonable endeavours to dissuade its other Subsidiaries from asserting their patents as aforesaid.

5.2   CD undertakes, on behalf of itself and its Named Subsidiaries, not to assert against AB present and future patents owned by CD and its Named Subsidiaries in all countries insofar as such patents are for inventions contained in, or exemplified by, Technical Information disclosed to AB hereunder, to prevent the development, manufacture and sale of lubricant additives. Additionally, CD will use reasonable endeavours to dissuade its other Subsidiaries from asserting their patents as aforesaid.

5.3   Neither party shall be deemed by the terms of this Agreement to have made any representation as to the validity of any patent referred to in 5.1 or 5.2 of this Article 5.

5.4   Subject to Article 12, the undertakings given in 5.1 and 5.2 of this Article 5, shall with respect to each patent remain in effect for the full life thereof.

5.5   CD and AB may extend the benefit of the undertakings in 5.1 and 5.2 of this Article 5 to any Subsidiary or Joint Venture. Each extension as aforesaid shall be notified to the other party in writing and shall confer an immunity only while the respective undertakings in 5.1 and 5.2 of this Article 5 continue in force in relation to the party extending its benefit. No other right or immunity under any patents is conferred by this Agreement.

ARTICLE 5: COMMENTS

5.1/5.2   A non-assertion undertaking. Named Subsidiaries patents committed. Note that only patents directed to the exchanged information are relevant but the freedom under those patents is suitably broad.

5.4   The undertakings cannot be rescinded or cancelled other than for uncured default (Article 12).

ARTICLE 6: PROCEDURES

6.1   Technical Information will be transferred hereunder by whatever means is convenient to the parties, including a reasonable number of visits by each to the facilities and plant of the other, exchange of reports, operating procedures and equipment construction details (each party bearing its own costs and expenses).

6.2   The parties will meet as soon as practicable after the effective date hereof to discuss and agree on (i) detailed procedures and timetables for exchanging Technical Information and (ii) safeguards to prevent the exchange procedures becoming unduly burdensome to either party.

6.3   It is envisaged that each party will indicate to the other in general terms the Technical Information available for exchange hereunder, and the other will then indicate which Technical Information it wishes to receive in detail. All UK and US patents and patent applications for inventions contained in or exemplified by Technical Information available to a party for exchange hereunder shall be copied by the party to the other at the earliest legally permitted time.

ARTICLE 6: COMMENTS

6.1   Schedule 2 relevant.

6.3   The last sentence is important in case the parties should decide to be even more selective about what they want to receive. Note that Article 5 concerns only the information disclosed, not what might have been disclosed. In the UK, for example, there are statutory restrictions on disclosing new patent filings until a security review period has passed. If a restriction order should be imposed, permission to disclose is required.

ARTICLE 7: TERM
Subject to the provisions of Articles 2.2, 9 and 12 this Agreement shall come into force on the date first above written and shall remain in force thereafter for 5 years. In the fifth year the parties will meet to discuss whether and on what terms this Agreement might be extended. The rights and undertakings set out in Articles 3, 4, 5 and 8 shall survive termination of the Agreement pursuant to this Article 7 or Article 2.2.

ARTICLE 7: COMMENTS
This clause determines the exchange period and is the back-stop Cut Off Date. Note that, apart from in a default situation, the benefits and the Secrecy obligations continue.

ARTICLE 8: SECRECY
Save as is herein specifically provided, each party will keep secret and confidential and will use only as herein provided Technical Information received from the other hereunder. Each party will endeavour to restrict the disclosure of such

received Technical Information to those of its directors, officers, agents and employees who are involved in the research, development, manufacture or sale of lubricant additives and will use its reasonable endeavours to ensure that they know of and observe the provisions hereof. Each party will also obtain from its Subsidiaries, Co-owners in Joint Ventures, and other parties to whom disclosures are made, written undertakings of equivalent effect to the provisions of this Article. The provisions of this Article shall remain in force in respect to each item of Technical Information received for a period of 15 years from the date of its receipt, but shall not apply to:

(i) information which is in or comes into the public domain through no fault of the recipient hereunder;

(ii) information which the recipient can show to the reasonable satisfaction of the other party hereto was in its possession prior to receipt hereunder;

(iii) information which the recipient receives from a third party with good legal title thereto to the extent that disclosure or use is permitted by such third party;

(iv) information that in the considered professional judgement of the recipient's patent attorneys/agents it is necessary to include as supporting disclosure in any patent specifications directed to an invention made wholly or in part by one or more inventors in the employ of the recipient, in order to meet legal requirements for adequacy of disclosure and patent validity.

ARTICLE 8: COMMENTS

Unexceptional save for item (iv). This exception was necessary so that patenting of new discoveries in a Core Business would not be constrained. Some of these patents, but by no means necessarily all, would be covered by Article 6.3.

ARTICLE 9

If, in the reasonable opinion of either party, the business circumstances of the other should change to a significant extent at any time during this Agreement by reason of actual or impending merger, bankruptcy, or sale of its Products manufacturing operations, then the party unaffected shall be entitled to call for an immediate review of the provisions of this Agreement and the parties agree promptly to negotiate such mutually acceptable amendments to this Agreement as may be reasonably necessary adequately to protect the interests of said unaffected party.

ARTICLE 9: COMMENTS
A comfort clause.

ARTICLE 10
Neither this Agreement nor any of the rights and obligations arising hereunder may be assigned by either party without the prior written consent of the other party and any attempted assignment in violation of this Article 10 shall be null and void and of no force or effect. The parties each agree that consent to assignment of the Agreement will not be unreasonably withheld wherever such assignment is conditioned upon and reasonably commensurate with a transfer of ownership of a party's entire business and assets for lubricant additives.

ARTICLE 10: COMMENTS
A necessary safeguard.

ARTICLE 11
Nothing in this Agreement shall be construed to restrict the manufacture, use or sale of any product by either party in any country of the world except as such manufacture, use, or sale may be in violation of the rights of the other party not licensed pursuant to this Agreement.

ARTICLE 11: COMMENTS
For the avoidance of doubt.

ARTICLE 12
Should either party be in default of any of its obligations hereunder and fail or be unable to remedy such default within 30 days of receiving notice thereof from the other party hereto, that other party may, by further notice in writing, terminate this agreement forthwith. In that event, Article 8 shall survive termination and, as regards solely the party not in default, the rights and undertakings in Articles 4 and 5 shall continue for its benefit and the benefit of its Subsidiaries and Joint Ventures.

This Agreement shall be governed and construed in all respects by the Law of England.

IN WITNESS whereof the parties hereto have executed this Agreement as of the day and year first above written.

AB COMPANY
BY: ....
TITLE: ....

CD INC
BY: ....
TITLE: ....

ARTICLE 12: COMMENTS
A necessary Article even though enforcement of such Articles can be a real headache. But if there is a clear major breach and there are relevant patents, the Article would have teeth.

SECTION 4: JOINT DEVELOPMENT AGREEMENT

This Agreement is made this ..... day of ....., 1984 by and between AB Company of .... and CD Company of ....

WHEREAS:

1. The parties (herein referred to as 'AB' and 'CD' respectively) have a common interest in the development of new synthetic building materials and each possesses considerable knowledge, skills and experience in the general field of synthetic building materials, their production and use.

2. The parties have identified a certain programme of work concerning the development of new synthetic building materials more resistant to rain and frost (herein called 'the Programme') which is appropriate for joint research and development and which promises, as a result of combining their skills, knowledge and efforts, a successful outcome in a relatively short period.

3. The parties have, accordingly, agreed to implement the Programme on the terms and conditions hereafter appearing.

NOW THEREFORE, the parties agree as follows:

ARTICLE 1: DEFINITIONS

1.1 'The Field' means cement products, concrete master batches and shape-determined structures (which may be reinforced) made therefrom.

1.2 'Programme' means the targets, methods, techniques, activities, studies, tests and evaluations determined for the joint research and development work hereunder in the Field, as further set out in the Schedule hereto and as that Schedule may be altered or extended by agreement.

1.3 The 'AB work programme' and the 'CD work programme' signify the complementary parts of the Programme carried out by AB and by CD hereunder as the case requires. As presently conceived, these work programmes are further specified in the Schedule.

1.4 'Input Information' means all technical information of either party which is relevant to the Programme and which that party has the unfettered right to use in its work programme or to disclose to the other party for use in that party's work programme.

1.5   'Received Information' means all such Input Information of either party as is disclosed to, or acquired by the other party pursuant to this Agreement.

1.6   'Developed Information' means all technical information (including, without limitation, all results, findings, discoveries and inventions) produced in the Programme.

1.7   'Development Inventions' means all patentable inventions included in or exemplified by Developed Information and conceived in the execution of the AB work programme and/or the CD work programme.

1.8   'Patent Rights' means patents, patent applications and similar intellectual property rights.

1.9   'New Patent Rights' means Patent Rights in respect of Development Inventions.

1.10   'Subsidiary or Affiliate' means any enterprise of which the party in question owns or controls at least 30% of the voting stock.

ARTICLE 2: THE PROGRAMMES

2.1   Each party shall provide to the other after the coming into force of this Agreement, its Input Information which would, in its considered judgement, be relevant to and materially assist the other party in the execution of that party's work programme. Such Input Information that is presently perceived to be relevant or of assistance shall be exchanged as soon as reasonably practicable and all such provision of Input Information shall be effected by such means and in such manner as the parties agree to be mutually convenient.

2.2   Each party shall use and draw upon in its execution of its work programme all its Input Information.

2.3   The obligations in Articles 2.1 and 2.2 apply to Input Information of either party that is possessed, in the case of AB by its Cement Division and in the case of CD by its Building Products Division.

2.4   Each party will use its reasonable endeavours to successfully execute and complete its work programme but neither party shall incur any liability to the other or be in breach of this Agreement should it, for any reason, fail to do so. If either party shall elect to discontinue its work programme or be prevented from continuing its work programme (other than temporarily) by reason of any lack of, or non-availability of, any resources required for its execution (whether due to circumstances under the control of that party or due to circumstances not under the control of that party), the party so electing or being so prevented shall give prompt notice thereof to the other and that other party shall have the right to terminate this Agreement by notice in writing.

2.5   The parties shall each be entitled to have and receive all Developed Information without prejudice to either party's right to withhold its secret, confidential or proprietary Input Information relating to the detailed means, methods, or techniques employed to produce such Development Information to the extent, but only to the extent, that such withholding would not prevent a sufficient and proper understanding of, and evaluation of, Developed Information by, or deny such Developed Information utility in the hands of, the receiving party. Having regard to the provisions of Article 5 (Secrecy), the parties do not intend to exercise this right in an overly protective or restricting manner.

2.6   All Developed Information shall be the property of each party independently to use and disclose freely and for all purposes subject only to Article 5 (Secrecy) and Article 6 (Patent Rights). Received Information shall be usable by a recipient party as set out in Article 4.

2.7   Whenever disclosures are made pursuant hereto of Developed Information in association with Received Information, the onus shall be on the disclosing party promptly to identify and distinguish all such Received Information and, except to the extent so identified and distinguished, all information disclosed which purports to be Developed Information shall, for all purposes of this Agreement and relating to it, be deemed to be Developed Information.

2.8   The parties shall each appoint a team leader who will have the day to day management of and control over its work programme and these team leaders shall be the principal persons through whom progress reviews of the work programmes shall be periodically carried out and any desired changes to the Programme or an individual work programme discussed and agreed upon.

ARTICLE 3: DEVELOPMENT INVENTIONS

3.1 AB shall have the first right to seek New Patent Rights in respect of any Development Invention conceived in the execution of the AB work programme in and for all countries.

3.2 CD shall have the first right to seek New Patent Rights in respect of any Development Invention conceived in the execution of the CD work programme in and for all countries.

3.3 The party not having such first right shall be entitled to require New Patent Rights to be sought at its cost and expense in any country or countries in respect of which the party having such first right elects not to seek New Patent Rights. The costs and expenses of filing, prosecuting and of grant and maintenance of New Patent Rights shall be borne by the party electing to seek, or requiring the other to seek, New Patent Rights. Neither party shall take any action, or fail to take any action, which results in any New Patent Rights lapsing or being withdrawn without first giving the other party the right to take over control and responsibility for those New Patent Rights.

3.4 The parties will establish appropriate consultation and communication between its patent agents and advisors as necessary to ensure that the provisions of this Article 3 are properly and effectively implemented to the benefit of both parties, and so that the rights of neither are impaired or prejudiced.

3.5 The parties recognise that their corporate policies on the acquisition of patents, as opposed to preserving inventions as confidential, secret and proprietary information, will not necessarily coincide in the case of any given Development Invention. Accordingly, each party will before exercising its rights herein take due note of any representations made by the other as regards patenting policy in general or in particular instances.

3.6 Should there be a Development Invention of which at least one inventor is employed by AB and at least one inventor is employed by CD the parties will discuss and agree what New Patent Rights will be sought, where, on what basis of costs and responsibility allocation.

3.7 All New Patent Rights shall be beneficially owned by both parties and each party shall have the right (irrespective of who is the applicant, registered proprietor, or legal owner) to make, use, exercise and vend Development Inventions under all such New Patent Rights in all countries, and have the right to license others, without accounting to the other party, save that neither party shall purport to grant any exclusive licence, or assign any title to or interest in New Patent Rights to a third party, without the consent of the other party hereto.

ARTICLE 4: USE OF RECEIVED INFORMATION

4.1   Subject to Articles 5 and 6, each party may use Received Information freely, without restriction, for all purposes, and may authorise any Subsidiary or Affiliate to use Received Information for any purpose.

ARTICLE 5: SECRECY

5.1   Neither party shall disclose any Development Invention to any third party (except under binder of secrecy) until such time as New Patent Rights have been sought therefor or until 3 years following its conception.

5.2   For a period of 10 years from the date hereof, neither party shall publish, or disclose to others, any Received Information excepting that:
(i) each party shall be entitled to disclose Received Information to the extent reasonably necessary (a) for its use, or best use, of Developed Information or (b) in relation to any research, development, production or use or sale of materials and products, for the purposes of the Programme;
(ii) each party shall be entitled to disclose Received Information to any Subsidiary or Affiliate accepting like terms, mutatis mutandis, as are contained in this Article 5.2;
(iii) each party shall be entitled to disclose Received Information in any specifications of, or applications for, Patent Rights in respect of Development Inventions or any other inventions made wholly or in part by that party, to the extent reasonably necessary under patent laws and practices to support or substantiate disclosures of the inventions;
provided always that Received Information shall be made subject to the same policies as are applied in the party in question for the due protection of its own confidential, secrecy or proprietary information of like character.

5.3   The obligations in Article 5.2 shall not apply to Received Information that corresponds in substance to:

(i) information in the public domain, or becoming so without default on the part of any recipient;

(ii) information already in the possession of the recipient at the time of receipt hereunder or pursuant hereto, which the recipient is free to disclose;

(iii) information received subsequently from an independent third party which the recipient is entitled, under his agreement or arrangements with the third party, to disclose.

ARTICLE 6: PATENT RIGHTS

6.1 No right, immunity, or licence is granted or implied in this Agreement in respect of any Patent Rights except as set out in Article 6.2.

6.2 Each party agrees for the benefit of the other party, its Subsidiaries and Affiliates, that it will not assert Patent Rights (if any) owned or controlled by it which are for inventions conceived heretofore and which, or the claims of which, cover or restrict the use of exchanged Received Information and/or Development Information in relation to the Field, and which would (apart from this non-assertion undertaking) be assertable to prevent the production, use or supply in, or export to, any country of items or materials in the Field. The parties and their Subsidiaries and Affiliates may extend the benefit hereof to their customers.

ARTICLE 7: GENERAL

7.1 Subject to Article 2.4, this agreement shall continue in full force and effect for an initial period of 2 years from the date hereof and, unless terminated by either party giving to the other not less than one year's notice in writing, shall continue thereafter from year to year. In any event, the parties shall formally review this Agreement towards the end of its first year and will exchange views on the desirability of its continuance beyond the second anniversary and to consider any desired amendments. Termination of this Agreement shall not affect the provisions of Articles 2.5, 3, 5 and 6 which shall continue in full force and effect on their terms.

7.2 This is an Agreement made in England and subject to English Law.

7.3 Notices hereunder shall be sent to the following addresses for the parties:

For AB: ....
and marked for the attention of: ....
For CD: ....
and marked for the attention of: ....
   IN WITNESS WHEREOF, the parties hereto have caused this Agreement to be signed by their duly authorised signatories as of the day, month and year first above written.
By: ....
By: ....

SECTION 4: COMMENTS
This is an Agreement of the kind described in the footnote to Chapter 4. It is a vertical joint development agreement between a maker of cement-type materials (AB) and a user of such materials to make building materials (CD).
   It is offered without further comment! The reader is invited to analyse it himself/herself and to consider the strengths and weaknesses of the formulae and procedures relied upon against the questions and issues raised in the footnote to Chapter 4.

SECTION 5: PROPRIETARY PRODUCT EVALUATION AGREEMENT

LETTER AGREEMENT
'Customer'
Dear Sirs,

EVALUATION OF PROPRIETARY PRODUCTS

From time to time, by arrangement between representatives of our two Companies, we will be supplying to you (Customer) samples of development products so that you may evaluate their suitability for use in particular application areas. We may also, in that context, provide to you, as appropriate, certain technical information relating to the products, such as components, composition, characteristics, properties and test performance results.

As development products they will not yet have reached the status of products that are on our sales range and which are sold (patents apart) without restriction on what may be done with them or how they may be used. Accordingly they are regarded by us as proprietary materials that should be supplied to others only against specific undertakings given in writing which restrict their analysis and use (outside of evaluation in permitted application areas) and which safeguard such related confidential information as may be directly provided by us or may be acquired in the course of permitted evaluations.

We, therefore, require your agreement to the following terms and conditions which shall govern the supply by us to you of development products (herein 'Products') and related technical information (herein 'Technical Information').

1. Supply to Customer of any Product for evaluation in any specified application area shall not imply an exclusive relationship with Customer as regards that product or that application area, except as shall be expressly agreed in writing.

2. A product shall be evaluated for use only in application area(s) agreed and recorded in writing.

3. No analysis which has the object or effect of revealing the nature or formula of ingredients or the composition of a Product shall be performed by or on behalf of Customer except as and to the extent agreed and recorded in writing.

4. The results of analyses performed on Product by or on behalf of Customer shall be deemed to have been supplied by us and be Technical Information for all purposes of this letter agreement.

5. No quantity of supplied Product shall be passed out of the possession and control of Customer except as we shall approve in writing, such approval to include approval of any third party involved and approval of the terms and conditions governing the passing of Product to him to the extent necessary to assure us of protection equivalent to that afforded hereby. Where the specific agreed evaluation necessarily involves conversion of Product so that its identity is lost, or dispersal of Product so that it cannot be practicably recovered or isolated, such approval shall be implicit in approval of the mode of evaluation.

6. All quantities of unconsumed Product remaining after evaluation shall be returned to us or destroyed as we may direct. Unless otherwise agreed, an evaluation shall be presumed completed after (12) months from first supply of Product therefor.

7. Technical Information shall be held by Customer in confidence and made available only within Customer's organisation and only to those persons who need to have it for the purposes of the evaluation, its execution and management. Customer may request authority to disclose Technical Information to a Parent, Subsidiary or Affiliated Company or to a Contractor or Consultant and we will not unreasonably withhold consent nor impose unreasonable conditions on such disclosure. We may require a secrecy agreement direct with the proposed disclosee or, alternatively, may require that the secrecy agreement between Customer and the disclosee expressly state that the undertakings given by disclosee are for our benefit.

8. Technical Information shall be used by Customer only for the agreed evaluation and shall not otherwise be used in any research or business activity.

9. Upon completion of Customer's evaluation, Customer shall deliver to us a written report which shall fairly and accurately reflect Customer's conclusions as to the suitability of Product for use in the application area evaluated and shall present the basic results on which those conclusions are founded. We may disclose and use those conclusions and results in our promotion of Products but shall not attribute them to Customer or Customer's operations as their source.

10. All results and findings of Customer's evaluation of Product shall be the property of Customer to use or disclose as Customer thinks fit. However, Customer shall not in any disclosure associate such results and findings with the use of our development products without our agreement to content and context. We reserve the right to be the sole source of promotional material or publicity concerning the utility of our development products.

11. If either party shall consider that the results of the evaluation provide a basis for worthwhile patent protection for an invention relating to a field of industrial application using products of a class of which Product is a member, and it is necessary in order to support or substantiate the relevant invention to disclose in the patent specification information that, but for the evaluation of Product by Customer, would not have been available, that party shall consult with the other, and the parties shall in good faith determine and agree a patenting strategy, and the ownership and licensing rights each party shall have, irrespective of inventorship. Neither party shall seek or acquire patent protection for such an invention other than in accordance with such agreement.

12. The supply and evaluation of Products pursuant to this letter agreement does not imply any obligation to supply or purchase commercial quantities of Products, nor any rights under any patents of either party.

13. Undertakings of non-disclosure and restrictions on use of Technical Information contained herein shall not prevent or restrict Customer from disclosing or using, as the case may be, information, however similar, that:
(i) is now, or subsequently without default becomes, in the public domain;
(ii) is already possessed by Customer as evidenced by written records;
(iii) is received by Customer from an independent source legally entitled to supply it, in so far as the terms of its receipt permit disclosure or use;
(iv) is subsequently independently developed by Customer by persons in its employ who have not had access to Technical Information or Product or, in any event, have not materially relied upon Technical Information to which the undertakings herein otherwise apply (or used Product) in the planning or execution of their development programme and activity, as Customer can reasonably demonstrate.

If these terms and conditions are acceptable to you, please have one of the duplicate originals of this letter signed on behalf of your Company and return it to us for our records, whereupon this letter will become a legal agreement between your company and us.

    Yours faithfully

ACCEPTED AND AGREED for and on behalf of: ....
BY: ....
TITLE: ....
Date: ....

SECTION 5: COMMENTS

This agreement provides a basis for customer evaluation of development products not yet released on to the open market. It does not involve collaborative joint development; merely supply of a quantity of the proprietary development product by its producer and evaluation of it by an interested customer for a defined use. It might lead on to a simple supply arrangement with or without preferred customer status, or to a joint development agreement if further development is shown to be required, or to technology licensing in either direction. Ideally, the evaluation needs to proceed in a manner that does not prejudice either party or close off options. In practice, the parties' interests are reasonably compatible until the product is shown to have a performance edge in a given application but from that point their interests conflict. This agreement is an attempt to manage that uneasy situation.

1. Freedom to have similar dealings with others reserved.

2. Use of Product restricted.

3. Analysis restricted (but the supplier would have to reveal enough to satisfy Health & Safety concerns).

4. Information derived from the Product deemed supplied.

5. Safe custody of Product.

6. Limited period for evaluation.

7. Secrecy of Technical Information (but see 13).

8. Use of Technical Information only for the specific evaluation (but see 13).

9. Customer reporting of broad conclusions and basis for those conclusions. Supplier free to use the report for Product promotions, but may not reveal source.

10. Customer can do what he likes with the results of his evaluation but may not say whose product was evaluated.

11. This is the crucial clause. Neither party can take advantage of this evaluation to patent an invention on a new use of such a Product. A 'but for' test governs, irrespective of considerations of confidentiality. A patent in either party's hands would prejudice the other, unless obtained by agreement. The usual position is that either party could prevent generalised patenting by openly disclosing what it knows, free of confidentiality. But probably neither party alone would have sufficient information at its disposal for a solid well-supported patent.

12. For the avoidance of doubt.

13. The necessary 'gateways', including independent development.

SECTION 6: TOTAL TECHNOLOGY LICENCE

LICENCE AGREEMENT

THIS AGREEMENT made and entered into at .... on this ..... day of .... 19..... by and between AB Company, a company organised and existing under the laws of England having its registered office at .... hereinafter referred to as LICENSOR, and CD Incorporated, a company organised and existing under the laws of Country Y having its registered office at .... hereafter referred to as OWNER.

WITNESSETH

WHEREAS LICENSOR is engaged in the production of Compound X by a process of .... (hereinafter referred to as PROCESS which is defined in more detail in Article 1 hereof) and possesses technical information, knowledge and experience required for the design, engineering, construction and operation of plant to produce Compound X by the PROCESS (hereinafter referred to as TECHNICAL INFORMATION which is defined in more detail in Article 1 hereof) and has the right to grant licences for the use of the PROCESS and TECHNICAL INFORMATION in Country Y;

WHEREAS OWNER is interested in the production and sale of Compound X in Country Y and desires to obtain the right and licence for use of TECHNICAL INFORMATION in a plant to be constructed by OWNER for the manufacture of Compound X in Country Y (hereinafter referred to as PLANT which is defined in more detail in Article 1 hereof).

WHEREAS LICENSOR is willing to disclose to OWNER and grant to OWNER the right and licence to use TECHNICAL INFORMATION as necessary for the purposes of design, engineering, construction and operation of the PLANT upon terms and conditions hereinafter set forth.

NOW THEREFORE, in consideration of the rights and obligations herein set forth, the parties hereto agree as follows:

COMMENTS

This is a total technology licence for the production in a specific plant of a compound here called Compound X to a company in a country with a centrally controlled economy (here called Country Y). It involves the Licensor in a major commitment of effort and resources for plant design, project planning and management (by secondment), detailed engineering design review, plant construction review, plant commissioning support, and plant operator training. The

145

Licensor is aided by a named International Process Plant Contractor (who was experienced in the supply of technology to Country Y).

Recitals: The first recital shows that this is a licence for a specific process and a specific parcel of technical information. It represents the Licensor's capability and right to licence.

The other recitals show that this is a single plant licence (although improvements, expansions and further plant are referred to in the Agreement).

ARTICLE 1: DEFINITIONS

The following terms as used herein shall have the following meanings for all purposes of this agreement:

1.1 'AGREEMENT' shall mean this Agreement of 25 Articles and includes all Schedules and Appendices hereto and all documents herein specified and any amendments which the parties may hereafter agree in writing to be made to this AGREEMENT.

1.2 'EFFECTIVE DATE' shall mean the date on which this AGREEMENT will come into force under the provisions of Article 15 hereof.

1.3 'IMPROVEMENTS' shall mean all modifications and refinements, patented or otherwise of the PROCESS which are developed by or for LICENSOR, or OWNER, as the case may be, after the EFFECTIVE DATE but within the period of exchange of IMPROVEMENTS as set forth in Article 3 below, being modifications and refinements brought to the status of commercial applicability and relevant to the PLANT and the PROCESS practised therein, and which one party may lawfully disclose to the other for use.

1.4 'PROCESS' shall mean the process owned by LICENSOR and employed at the EFFECTIVE DATE in commercial plant in the United Kingdom owned by LICENSOR for the production of Compound X of quality suitable for direct reaction with polyols to produce urethane polymers, which process comprises a first stage in which .... (feedstock) is reacted with .... to produce crude Compound X and a second stage in which this crude material is purified by ....

1.5 'PATENTS' shall mean any or all of patents relating to the PROCESS owned by LICENSOR which, or the claims of which, cover inventions contained in or exemplified by TECHNICAL INFORMATION or IMPROVEMENTS.

1.6 'TECHNICAL INFORMATION' shall mean information, knowledge and experience relevant to the PROCESS in the possession of LICENSOR at the EFFECTIVE DATE which is in LICENSOR'S judgement sufficient to enable an engineering organisation skilled in the engineering and construction of chemical plants in Country Y to complete the engineering of, and to construct, the PLANT and to enable OWNER to operate the PLANT so as to provide Compound X of merchantable quality, and includes information and designs in the DESIGN MANUALS and LICENSOR'S IMPROVEMENTS.

1.7 'PLANT' shall mean the plant to be constructed by and for OWNER in Country Y employing the PROCESS with an initial annual design capacity, based on 7,700 operating hours per year of 10,000 metric tonnes per annum of Compound X. This is more fully described in Appendix A.

1.8 'EXPANSION' shall mean increase in the annual capacity of the PLANT achieved by additions to, or duplication or major modification of, the main plant items and equipment and to be measured as the parties shall determine following DATE OF START-UP and after any Performance tests under Article 8 are completed.

1.9 'DATE OF START-UP' shall mean the date on which the PLANT shall have produced 10 tes of purified Compound X.

1.10 'CONTRACTOR' shall mean one or more Engineering Contractors appointed by OWNER to complete the engineering of and construct the PLANT.

1.11 'DESIGN MANUALS' shall mean designs, drawings, specifications and technical data of agreed content, definition and scope which are to be produced by LICENSOR from the BASIS OF DESIGN for OWNER's use for completion and construction of the PLANT. The requirements of the DESIGN MANUALS are set out in Appendix B.

1.12 'BASIS OF DESIGN' means the requirement for the process design of the PLANT to be developed and agreed between OWNER and LICENSOR. Preliminary requirements for the BASIS OF DESIGN are set out in Appendix B hereto.

1.13 'SERVICES' means advisory and training services provided by LICENSOR pursuant to Article 5.

1.14 'MAN DAY' shall mean any Calendar day or part thereof during which an employee of LICENSOR is deputed to furnish SERVICES in Country Y pursuant to this Agreement including days spent in travel between the place where such work is to be performed and the employee's home base in LICENSOR's operations.

1.15 'MECHANICAL COMPLETION' shall mean in relation to the PLANT the completion of installation of all equipment, piping and instrumentation including hydrotesting and/or pneumatic testing and other treatments as may be required and calibration of instruments and relays, so that the PLANT is prepared to receive PROCESS feedstock.

ARTICLE 1: COMMENTS

1.1 The Appendices are fundamental and three of them (B, C and D) are subordinate agreements.

1.2 Article 15 is essential for a licence to a controlled economy.

1.3 Note that this concerns only improvements that are commercially proven and applicable (or applied) to the Plant and its expression of the Process.

1.4 This shows the Process is the Licensor's current technology and characterises it by its important stages.

1.5 Note that this only includes patents relating to the production process and directed to patented features of the supplied technology. It does not, in particular, include patents for any downstream uses of Compound X.

1.6 This establishes the quality of information to be supplied. The Licensor is not to be concerned with detailed mechanical, electrical and civil engineering or with facilities outside 'battery limits', eg utilities supply. It sets a subjective standard (but it has to be) and so, pervading this licence, is the Licensor's public reputation; something the licensee and his Government will have taken into account.

1.7 Appendix A is obviously a key document.

1.8 An uneasy definition but reasonable and the best the parties could do at the outset.

1.9   Relevant to payments and performance tests. At this date the Plant will clearly be basically operable.

1.11/1.12 A lot of effort by the parties and their Contractors will go in to the specification of the requirements for the Design Manuals. The Basis of Design was yet to be decided in detail. The Assistance Agreement (Appendix C) provided for a secondment to the Licensee to help him get this decided and to plan his Project. This has proved to be an enormous help for all concerned when the Licensee is inexperienced in major projects and an intelligent beginner so far as the technology is concerned.

1.15   This is an important Project stage but only marginally relevant to the Agreement (Article 8.2).

ARTICLE 2: GRANT OF LICENCE

2.1   LICENSOR agrees to grant and hereby grants to OWNER as of the EFFECTIVE DATE, a non-exclusive right and licence under TECHNICAL INFORMATION:
(i) to complete the engineering design of, construct and operate the PLANT and EXPANSION(S) thereof for the manufacture of Compound X;
(ii) to sell or use in Country Y and to export, sell, or use in any other country Compound X so manufactured;
(iii) to purchase, acquire, make, or have made any equipment, apparatus, catalyst, chemical or other material necessary for the purposes set out in (i) and (ii).

2.2   LICENSOR, being the proprietor of know-how relevant to the manufacture of catalysts used by LICENSOR in the PROCESS, agrees to grant and hereby grants to OWNER a licence for the manufacture in Country Y of such catalysts for use in any plant operated by OWNER in Country Y and to disclose to OWNER such know-how in LICENSOR's possession at the EFFECTIVE DATE. This know-how will form a section of the DESIGN MANUALS.

2.3   LICENSOR undertakes not to assert against OWNER any Patent owned by LICENSOR which could be asserted but for this undertaking to prevent OWNER, his contractors, servants or agents, doing any of the acts recited in Article 2.1 (i) and (iii) and Article 2.2 or to prevent OWNER, his agents, or customers selling in and to any country listed in Schedule 1 hereto

PRODUCT produced by OWNER in the PLANT. Save as expressly recited in this Article 2.3, no right or licence under any patents of LICENSOR is granted hereby.

2.4   If OWNER shall at any time during the life of this Agreement request in writing a right to use TECHNICAL INFORMATION supplied pursuant to this Agreement for the purposes of the design, engineering, construction and operation of other plant in Country Y to produce PRODUCT, LICENSOR will not unreasonably refuse to extend to OWNER the necessary consents and licences for that purpose, nor will LICENSOR make such extension subject to unreasonable terms, conditions, or licence fees. Except, however, as may be agreed in writing in respect of any such extension LICENSOR shall not be required to perform any services or extend any warranties in connection with such additional plant.

ARTICLE 2: COMMENTS

2.1   This is a declaratory of what the Technical Information is supplied for (but see Article 14.2). Patents are considered in 2.3.

2.2   This was a special ancillary supply of know-how.

2.3   This is the patent non-assertion undertaking. Patents for the Process and aspects of the Plant were not, in the circumstances of the real antecedent agreement, very significant but were of some value to the Licensor to stop exports to his important markets. The exclusion of Patents for downstream uses of Compound X was, however, very important.

2.4   A statement of intent whose performance will be judged in the forum of public opinion.

ARTICLE 3: MPROVEMENTS

3.1   LICENSOR undertakes to disclose his IMPROVEMENTS to OWNER arising in the period ending on the 5th anniversary of the EFFECTIVE DATE free of any payment and OWNER shall have the right and licence to use such IMPROVEMENTS as TECHNICAL INFORMATION free of licence fees.

3.2   OWNER shall disclose to LICENSOR his IMPROVEMENTS arising in the period specified in Article 3.1 free of any payment and LICENSOR shall have and is hereby granted by OWNER a non-exclusive right and licence under such IMPROVEMENTS free of licence fees:
(i) to use and practise such IMPROVEMENTS in its facilities wherever located, and
(ii) to sub-licence such IMPROVEMENTS to his other licensees in all countries of the PROCESS, with freedom from suit under any patent acquired by OWNER or any assignee of OWNER directed to the use of such IMPROVEMENT in the PROCESS.

3.3   The foregoing disclosure obligations in relation to IMPROVEMENTS shall not imply a duty to produce information in a form specifically adapted to the other party's requirements free of charge.

3.4   LICENSOR undertakes to negotiate in good faith with OWNER, upon request, to establish a mutually acceptable basis for extension of the period of exchange of IMPROVEMENTS so as to encompass IMPROVEMENTS arising prior to the 5th anniversary of START-UP of the PLANT or, if earlier, the 10th anniversary of the EFFECTIVE DATE.

ARTICLE 3: COMMENTS

3.1   Important but would very likely complicate determination of Expansion fees.

3.2   A comprehensive grant-back provision (Note 5.1(iv), last sentence).

ARTICLE 4: DESIGN MANUALS

4.1   As soon as practicable following the EFFECTIVE DATE, OWNER and LICENSOR shall finalise the BASIS of DESIGN, whereupon LICENSOR shall commence production of the DESIGN MANUALS to the requirements and on the terms set out in Appendix B (Design Agreement). LICENSOR will be engaging the assistance of Messrs .... of .... in the production of the DESIGN MANUALS and in relation to certain of the SERVICES described in Article 5.

4.2   Appendix B includes a programme for the production and delivery of discrete portions of the DESIGN MANUALS and LICENSOR shall use all reasonable endeavours to adhere to such programme. However, any reference to dates and to periods of elapsed time are best estimates and targets as at the EFFECTIVE DATE; they are not to be construed as of the essence of this AGREEMENT.

4.3   LICENSOR shall keep OWNER fully and promptly informed of any actual or expected deviations from the programme set out in Appendix B and will consult with OWNER on any reasonable measures that might advisedly be taken to regain conformity with that programme or on the most suitable adjustments to the programme that might be made.

4.4   OWNER shall take delivery of DESIGN MANUALS at ..... (non Country Y location).

ARTICLE 4: COMMENTS

4.1 – 4.3   Important but self-explanatory.

4.4   Principally relevant to withholding taxes on payments under the Design Agreement (see 2.5 and 3.3 of the Design Agreement).

ARTICLE 5: SERVICES

5.1   LICENSOR shall provide the following SERVICES:
(i) Allow representatives of OWNER and/or CONTRACTOR to visit the PROCESS plant of LICENSOR in the UK. The total duration of such visits is not expected to exceed 10 days and shall not involve more than 6 representatives of OWNER and/or CONTRACTOR and will be scheduled so as not unduly to interfere with the operation of such plant.
(ii) Pursuant to the Design Agreement (Appendix B hereto) review and approve, in the UK, information, documents and drawings as specified in a Schedule thereto which have been prepared by OWNER or the CONTRACTOR in relation to completion and construction of the PLANT.
(iii) Provide for OWNER'S personnel practical training in operation, material handling, maintenance, quality control, effluent treatment, pollution control, and safety at LICENSOR'S plant in the United Kingdom employing the PROCESS on the terms and conditions set out in Appendix D (Training Agreement).

The training programme shall be conducted in the English language. All expenses of and responsibility for OWNER'S employees during the period of training shall to OWNER'S account.

(iv) Pursuant to Appendix C hereto (Assistance Agreement) make personnel available at the site of the PLANT during construction, commissioning and initial operation of the PLANT. Thereafter, LICENSOR will make available to OWNER qualified personnel in such numbers as LICENSOR shall determine after discussion with OWNER for one trip per year of up to 15 man-days, until the third anniversary of the DATE OF START-UP to consult with and advise OWNER with respect to any difficulties encountered by him in the operation of the PLANT using TECHNICAL INFORMATION. OWNER shall bear the cost of transportation to and from Country Y and of living accommodation in Country Y, in accordance with the provisions of Article 7 below. In any event, LICENSOR shall have the right at LICENSOR'S cost and expense to send representatives to visit the PLANT, witness PLANT operations and inspect PLANT operating records once per calendar year during the period of exchange of IMPROVEMENTS.

ARTICLE 5: COMMENTS
Self-explanatory. Note that the design reviews are to be held in the UK (tax considerations, as well as convenience to Licensor and his contractor). Note also the last sentence of 5.1(iv). This bolsters the obligation in 3.2.

ARTICLE 6: LANGUAGE
The disclosure of information under or pursuant to this Agreement including all manuals shall be in the English language.

ARTICLE 6: COMMENTS
Innocent, but in fact very important for the Training Agreement (Appendix D).

ARTICLE 7: TRAVEL AND LIVING EXPENSES

7.1 OWNER shall provide or, where applicable, reimburse the cost of the following facilities to LICENSOR's personnel assigned to perform SERVICES in Country Y under Article 5 above, for their travel to and from Country Y, their living, accommodation and transport expenses in Country Y:

(i) The price of a round trip economy class air ticket from the normal place of assignment of such personnel to Country Y;

(ii) Travel by air-conditioned first class train accommodation to the place of work on arrival in Country Y and from their place of work on departure on termination of the assignment;

(iii) While travelling in Country Y on work assignments, cost of travel by air-conditioned first class train accommodation or by air, and lodging in a suitable hotel;

(iv) Appropriate residential accommodation while on assignment at the location of the PLANT. Requirements will be discussed and agreed having regard to availability of accommodation and the duration of individual periods of assignment;

(v) Transportation facilities to and from the place of residence and the place of work;

(vi) Office accommodation, clerical and secretarial assistance and expenses of facsimile, telegram and telex messages and telephone calls required for the work;

(vii) Living expenses. OWNER shall pay LICENSOR'S personnel a living expense allowance of .... per day per person (net of local income taxes) for the period of such assignment in Country Y.

7.2   Local income tax payable on allowances to LICENSOR'S assigned personnel shall be borne and paid by OWNER. OWNER shall obtain and forward to LICENSOR or LICENSOR'S personnel as the case may be receipts in evidence of such payment.

ARTICLE 7: COMMENTS
Very important detail, especially to the personnel concerned! Tax advice, too, is very important when drafting this sort of clause.

ARTICLE 8: PROCESS GUARANTEES

8.1   LICENSOR guarantees that during a performance test conducted in accordance with the reasonable advice of LICENSOR'S representatives, the PLANT shall produce Compound X which meets the specification indicated in Appendix E at production rates stipulated in Appendix E, provided that all raw materials, catalysts, chemicals and utilities shall have the specifications stipulated in Appendix E. Additionally LICENSOR shall offer a guarantee of aggregate utilities consumption when the engineering design of the Plant is sufficiently developed. LICENSOR shall be deemed to have satisfied its obligations under its guarantee when during any one performance test lasting for a continuous period of 72 (seventy two) hours the PLANT shall have demonstrated the production of .... MT of Compound X meeting such specification with aggregate consumption of utilities as so specified:

(i) provided that the PLANT is competently engineered and constructed in conformity with the DESIGN MANUALS and is operated and maintained in accordance with procedures approved by LICENSOR (such approval not be unreasonably withheld);

(ii) provided further that the specifications of raw materials, chemicals, catalysts and utilities supplied to the PLANT shall have been measured by analytical and control methods approved by LICENSOR (such approval not to be unreasonably withheld).

8.2   Commissioning and Performance Tests. Following MECHANICAL COMPLETION and upon receipt by LICENSOR of OWNER'S notice of the expected date when feedstock is to be introduced into the PLANT, LICENSOR shall make available qualified personnel as stipulated in Article 5 hereof who shall be present to witness and advise on the commissioning and initial operation of the PLANT and any Performance Tests, after joint confirmation by LICENSOR, OWNER and CONTRACTOR that PLANT appears to be constructed in accordance with the DESIGN MANUALS. A date of Performance Test shall be mutually decided and agreed to by above parties concerned, which date shall be as soon as reasonably practicable after the DATE OF START-UP. In the event of the first performance test not fulfilling the guarantees referred to in Article 8.1 hereof, the test shall be repeated as often as may be required by LICENSOR but not exceeding ..... tests within ..... months from the DATE OF START-UP, provided that OWNER shall not be required to make any payment to LICENSOR towards provision of service of personnel for any period beyond the second test run or two months from the DATE OF START UP whichever shall be earlier, unless their period of assignment is extended by reasons not attributable to LICENSOR.

Where instruments are to be used they shall be calibrated before tests and if considered to be inaccurate, recalibrated after tests and an agreed instrument inaccuracy established. Instrument tolerances shall be agreed between OWNER and the LICENSOR before the guarantee test runs.

8.3   Should the PLANT, during such test run/runs not meet with the requirements for fulfilling the guarantees stipulated in this Article, LICENSOR shall recommend the design changes which are necessary to meet the requirements for fulfilment of the guarantees and all expenses incurred on such changes, where due to errors in the DESIGN MANUALS, shall be borne by the LICENSOR up to the LIMIT OF LIABILITY stipulated in Article 22.

8.4   If in no test run carried out hereunder is the performance of the PLANT such that the guarantees are met, and the reason or predominant cause is a deficiency in LICENSOR'S performance, LICENSOR shall pay to OWNER by way of compensation an amount arrived at in the manner indicated in Appendix F.

8.5   In the following cases LICENSOR shall be deemed to have fulfilled its guarantee obligations:
(i)   if performance testing cannot be started within 12 (twelve) months from DATE OF START-UP or 60 (sixty) months as from the EFFECTIVE DATE hereof, whichever comes earlier through no fault of LICENSOR, or
(ii) if performance guarantees cannot be attained within 24 (twenty four) months from DATE OF START-UP or within 12 (twelve) months from the start of the first performance test, whichever shall be earlier, through no fault of LICENSOR, or
(iii) OWNER elects not to have a performance test.

ARTICLE 8: COMMENTS
This clause should be reviewed against the relevant sections of the main body of this book. 8.5 is a key safeguard for the Licensor.

ARTICLE 9: PRICE AND TERMS OF PAYMENT

9.1   In consideration of the rights and licence herein granted to practise and use TECHNICAL INFORMATION in and for the PLANT, OWNER shall pay to LICENSOR a licence fee of ..... (..........).

9.2   Payment shall be made in the following instalments:
(i)   One third of the amount stipulated in Article 9.1 shall be paid by OWNER within 30 days of the EFFECTIVE DATE;
(ii) One third of the amount stipulated in Article 9.1 shall be paid within 30 days of the date on which the DESIGN MANUALS are delivered to OWNER pursuant to Appendix B;

(iii) One third of the amount stipulated in Article 9.1 shall be paid within 30 days of the DATE OF START-UP or, if sooner, the 4th anniversary of the EFFECTIVE DATE.

9.3 Expansion fees. OWNER shall pay to LICENSOR the following fee for any EXPANSION of the PLANT achieved during the terms of this Agreement, ... per metric tonne of additional annual capacity, payable within 60 days of the end of the calendar year in which the EXPANSION is achieved.

ARTICLE 9: COMMENTS
Discussed sufficiently already.

ARTICLE 10: TAX
All taxes, levies and other similar payments payable outside the United Kingdom on sums due to LICENSOR under this Agreement shall be borne and paid by OWNER except to the extent that credit may be obtained for such taxes, levies or other payments against United Kingdom tax payable on such sums, either:
(i) under any Double Taxation Convention between the Government of the United Kingdom and the Government of Country Y, or
(ii) under United Kingdom legislation for granting credit unilaterally.

ARTICLE 10: COMMENTS
Discussed sufficiently already.

ARTICLE 11: INFRINGEMENT OF PATENTS

11.1 Patent Warranties. LICENSOR warrants that construction and operation of the PLANT in accordance with the DESIGN MANUALS and sale in Country Y of Compound X so produced will not cause OWNER to infringe any patent owned by third parties as of the EFFECTIVE DATE.

11.2 Defence of claims and suits. In the event that any claim is made or suit or other action is brought against OWNER, upon the basis that the PLANT or its operation by OWNER and/or the sale of Compound X produced thereby infringes a patent of a third party, OWNER shall promptly notify LICENSOR of such claim, suit or action and LICENSOR shall provide advice to assist OWNER to answer such claim or defend such suit or action.

Except where deriving from a breach of the warranty in Article 11.1, OWNER will reimburse LICENSOR'S reasonable costs and expenses.

ARTICLE 11: COMMENTS

11.1   Note that this warranty only relates to third party patents in Country Y. If the technology is new and being competitively developed by others, some provision would have to be made for unpublished pending patent applications.

11.2   The Licensor assists the Licensee to defend claims for patent infringement.

ARTICLE 12: INSURANCE
Both parties will effect insurance for their own personnel and property respectively. Particularly, LICENSOR will at its expense place insurance to cover any claim that may be made alleging bodily injury or death sustained by LICENSOR'S employees as a result of or in connection with the performance of this Agreement, and will hold OWNER and its employees and representatives harmless from any and all of such claims.

LICENSOR'S insurance policy as to its property and personnel shall include a waiver clause as to any insurer's action against OWNER. OWNER will at its expense place insurance to cover any claim that may be made as a result of bodily injury or death to any of OWNER's employees as a result of or in connection with LICENSOR'S performance of this Agreement and will hold LICENSOR harmless from any and all of such claims. OWNER's insurance policy as to its personnel and property shall include a waiver clause as to any insurer's action against LICENSOR.

ARTICLE 12: COMMENTS
This is an important clause and expert advice and insurers' cooperation is vital.

ARTICLE 13: MOST FAVOURED LICENSEE
If, subsequent to the EFFECTIVE DATE and during the term of this AGREEMENT, LICENSOR shall grant a licence on terms and conditions substantially similar to this licence to any unconnected third party in Country Y on a licence and know-how fee basis more favourable to said third party licensee than the terms specified in Article 9.3 for EXPANSIONS, LICENSOR shall promptly inform OWNER and OWNER shall be entitled to substitute such more favourable terms for the terms specified in Article 9.3.

ARTICLE 13: COMMENTS
See Chapter 4 of the main text.

ARTICLE 14: CONFIDENTIALITY

14.1 OWNER agrees until the 15th anniversary of the EFFECTIVE DATE to keep confidential and not to disclose to any third party any TECHNICAL INFORMATION received from LICENSOR pursuant to this AGREEMENT except such information as now is or hereafter becomes published or otherwise generally available to the public. OWNER is, however, authorised to disclose such information to his employees and to suppliers of equipment and engineering, construction and insurance companies in connection with the PLANT (subject to their giving in writing undertakings on terms approved by LICENSOR not to disclose such information to others and not to use it for purposes other than those for which it is to be disclosed to them).

14.2 OWNER agrees not to use commercially, except as expressly licensed hereunder or pursuant to Article 2.4, TECHNICAL INFORMATION to which the obligation in Article 14.1 apply.

14.3 OWNER is hereby authorised to transfer any technical information or data furnished by LICENSOR to any Ministry or Department of the Government of Country Y subject, however, to their first giving an undertaking in writing to be bound by the provisions of this Article to the same extent as OWNER.

ARTICLE 14: COMMENTS

14.1 Note the 15 year term and that the only 'gateway' is public knowledge. Note also Licensor approval of terms for onward necessary disclosures.

14.2 A vital supplement to Clauses 2. (See Appendix II of this book.)

14.3 This is a practical necessity in many licensing situations (and not only in controlled economies) for investment approvals, planning permissions, under building regulations and health and safety responsibilities, etc.

ARTICLE 15: EFFECTIVE DATE
The EFFECTIVE DATE of this Agreement shall be the date on which the conditions stated below have been fulfilled:

(i)   This Agreement is executed by both parties; and

(ii) OWNER notifies LICENSOR that all necessary approvals of the Government of Country Y (including the Reserve Bank) to the terms and conditions of this AGREEMENT have been received so that OWNER is legally and irrevocably bound by such terms and conditions and is authorised to remit all sums payable hereunder.

ARTICLE 15: COMMENTS

An essential safeguard, for both parties but especially the Licensor. See also the last sentence of Article 20.

ARTICLE 16: TERM

The term of this Agreement shall be 15 years as from the EFFECTIVE DATE. The rights and licences granted by LICENSOR to OWNER shall survive the expiration of this AGREEMENT by effluxion of time.

ARTICLE 16: COMMENTS

Nothing peculiar here.

ARTICLE 17: TERMINATION

Termination of this Agreement shall not relieve either party of any obligations then accrued hereunder or which on their terms extend beyond the date of such termination.

ARTICLE 17: COMMENTS

Nothing peculiar here.

ARTICLE 18: ASSIGNMENT

This Agreement shall not be assignable by either party without the prior written consent of the other party hereto, except that this Agreement shall be automatically binding upon, and inure to the benefit of, any successor of OWNER or any person, firm or corporation acquiring all or substantially all of the relevant business and assets of OWNER.

ARTICLE 18: COMMENTS

Nothing peculiar here.

ARTICLE 19: ARBITRATION

19.1 All disputes between the parties arising out of the provisions of this Agreement, which the parties do not promptly resolve amicably by negotiation and which they agree shall be settled by arbitration, shall be submitted for arbitration under the Rules of Conciliation and Arbitration of the International Chamber of Commerce.

19.2 The award of the Arbitrator or Arbitrators, as the case may be, shall be final and binding on the parties hereto. Judgement upon the award may be entered in any court of competent jurisdiction.

19.3 The arbitration proceedings shall be in the English language and shall be held in ....

ARTICLE 19: COMMENTS

19.1 Arbitration by agreement when an issue has arisen.

19.3 Article 23 applies the law of Country Y to this Agreement (often the Government of Country Y, whose approvals are required for the Agreement to take effect, will insist that local law apply). The place of arbitration may have to be Country Y also but if there is a neutral territory where matters of this sort affecting Country Y have previously been effectively dealt with (and respected by the authorities and judiciary in Country Y) that would be a preferred venue for the Licensor.

ARTICLE 20: FORCE MAJEURE

Notwithstanding anything in this Agreement to the contrary, neither LICENSOR nor OWNER shall be liable to the other, or deemed to be in default, if performance hereunder is prevented or delayed by Force Majeure which term shall mean fire, explosion, action of the elements, act of any Government, or other cause beyond the reasonable control of the party affected, provided that no such cause shall be deemed to be Force Majeure unless the party affected shall notify the other party within 20 days as from the beginning of operation of the said cause, and shall thereafter use all reasonable endeavours to overcome such cause of delay and resume performance. The affected party shall supply to the other party documents substantiating the cause of delay. If Force Majeure shall be invoked to excuse non-payment by OWNER of sums due under Articles 7 or 9, LICENSOR shall be entitled forthwith to suspend further performance hereunder until all overdue sums are paid.

ARTICLE 20: COMMENTS
The last sentence is something Licensors should insist on.

ARTICLE 21: NO WAIVER OF RIGHTS
The failure of either party hereto at any time to enforce any of the terms, provisions, or conditions of this Agreement shall not be construed as a waiver of the same or of the rights of either party to enforce the same on any subsequent occasion.

ARTICLE 21: COMMENTS
A common legally motivated clause.

ARTICLE 22: OVERALL LIMIT OF LICENSOR'S LIABILITY
Except as expressly stated otherwise in the agreements forming the Appendices hereto, the total liability of LICENSOR to OWNER for breach of any term or condition hereof, breach of warranty, payments pursuant to Article 8, and loss or damage occasioned by any act or omission of LICENSOR, his servants, or agents in or relating to the performance hereof, whether founded in contract, tort, strict liability or any other legal characterisation whatsoever shall be limited to ..... % of fees specified in Article 9.1 and such liability shall be settled by first applying a credit against any portion of such fees not yet paid pursuant to Article 9.2

ARTICLE 22: COMMENTS
This is a critical safeguard to the Licensor, whatever the ceiling of liability may be set at. Prior legal advice from Country Y on its effectiveness is essential.
    Increasingly, in agreements with US Licensors, the contractual limits (or exclusions) of liability will be highlighted, eg by being capitalised and even, sometimes, marginally noted by the parties' legal counsel to show that due advice was taken. In the USA, licensing agreements are commercial contracts and are governed by the relevant State Law and influenced by 'good practices' in that State, but marginal noting by lawyers seems excessive.
    It is inconceivable that parties would draw up and execute this sort of agreement without comprehensive legal/tax/insurance advice.

ARTICLE 23: INTERPRETATION

This AGREEMENT shall be governed by, be subject to and interpreted in accordance with the laws of Country Y. Any amendment or modification of this agreement shall be in writing and signed by a duly authorised representative of each party, and shall take effect when any necessary approvals of the Government of Country Y have been obtained.

ARTICLE 23: COMMENTS

Note that amendments have to be treated with all the formalities that apply to this Agreement.

ARTICLE 24: APPENDICES

Appendices A to F hereto shall form part of this AGREEMENT.

ARTICLE 25: NOTICES

All notices, reports, requests or demands to be given by either party to the other under the provisions of this Agreement shall be forwarded by registered airmail, or by cablegram or telex (and confirmed by registered airmail), properly addressed to the respective parties as follows:

LICENSOR: REPRESENTATIVE OF LICENSOR
                LICENSOR'S ADDRESS
OWNER: REPRESENTATIVE OF OWNER
                OWNER'S ADDRESS

or at such other address or addresses as either party may from time to time designate by written notice as its address or addresses for the purpose hereof.

IN WITNESS WHEREOF, the parties hereto caused this Agreement to be duly executed as of the day and year first above written.

OWNER

LICENSOR

SCHEDULE 1

APPENDIX A: Description of the PLANT

APPENDIX B: Design Agreement

APPENDIX C: Assistance Agreement

APPENDIX D: Training Agreement

APPENDIX E: Specifications and Guarantees

APPENDIX F: Compensation

APPENDIX B: DESIGN AGREEMENT

This Agreement is made this ..... day of .... 19..... by and between AB Company, a company organised and existing under the laws of England, of ...., London,(herein called 'AB')

and

CD Incorporated, a company organised and existing under the laws of Country Y of ..... (herein called 'CD'), and is supplemental to an Agreement (herein called the 'Licence Agreement') made by the parties hereto on the date first written above.

WHEREAS, under the Licence Agreement, AB undertakes to provide design data, drawings, information and specifications in the form of DESIGN MANUALS for a plant (therein and herein called the PLANT) to produce ..... (etc)

and

WHEREAS, the Licence Agreement stipulates that the form, content, scope and requirements of such DESIGN MANUALS and the terms for their supply shall be set out in a separate agreement which shall be annexed to the Licence Agreement as APPENDIX B.

NOW, THEREFORE, the parties have agreed the following as the terms and conditions of such separate agreement.

COMMENTS

The Design Agreement is largely self-explanatory. Only specially significant points are commented on below.

ARTICLE 1

This Agreement together with Schedules 1 and 2 hereto shall constitute APPENDIX B to the Licence Agreement.

ARTICLE 2

2.1 AB will supply to CD, for its use for completion, construction, and operation of the PLANT (as defined in the Licence Agreement) DESIGN MANUALS which shall be sufficient, in AB's considered judgement, to enable persons who are competent in the field of engineering, constructing and operating chemical process plants to build and operate the PLANT. It is presumed that CD will engage experienced Contractors to complete the engineering of, and construct, the PLANT for and on its behalf.

2.2　The form, content, scope and requirements of the DESIGN MANUALS aforesaid shall be as stipulated in Schedule 1, which includes CD's preliminary basis for the design of the PLANT. The definitive agreed basis for the design of the PLANT shall be finalised with all practicable speed by CD, in consultation with AB and in any event before ....

2.3　AB will be engaging the services of Messrs .... to assist in the preparation and production of the DESIGN MANUALS under separate sub-contract arrangements, without affecting AB's primary obligations and responsibility in relation thereto.

2.4　Schedule 1 includes a programme which displays the dates by which discrete portions of the DESIGN MANUALS can reasonably be expected to be available for delivery to CD, or to CD's nominated Contractor. The dates so displayed are target dates and, while AB undertakes to use all reasonable endeavours to procure that this programme is adhered to, AB does not guarantee adherence to such programme. AB will keep CD fully informed of the progress of the preparation of the DESIGN MANUALS, relating such progress to the programme, and will, when slippage occurs or is considered likely to occur, discuss with CD those measures which might advisedly be taken either to regain conformity with the programme or, if that is not practicable or required by CD, to amend the programme accordingly.

2.5　CD shall take delivery of the contents of the DESIGN MANUALS in the UK at a time or times and place or places to be agreed.

2.6　The DESIGN MANUALS shall be supplied to CD or CD's nominee Contractor in ..... copies in the English language with all units and measurements being according to the metric system.

2.7　AB will review and approve in the UK the information, documents and drawings specified in Schedule 2 hereto which have been prepared by CD or by CD's Contractor. The purpose of this review will be to ensure their conformity with the DESIGN MANUALS and approval will not extend to approval of general engineering aspects. AB's approval may be given orally or in writing (eg by fax or telex) but if given orally will be confirmed promptly in writing. This review process will be a substantial activity and AB is willing (in the interests of achieving efficient and timely reviews) to establish a continuous presence in the UK office of CD's Contractor throughout the relevant period of detailed engineering work.

ARTICLE 2: COMMENTS

2.1   Discussed under 1.6 of the Licence Agreement.

2.2   Article 1 of the Assistance Agreement is pertinent.

2.3   Both parties were using experienced process contractors (see 2.4, 2.6 and 2.7).

2.4   Reasonable endeavours. Note that basis of design is still to be agreed. Time for delivery of Design Manuals is not of the essence of this Contract.

2.5   CD's contractor had a UK Office but this clause had a tax significance.

2.7   This is an activity that both parties and their Contractors will want done thoroughly.

ARTICLE 3

3.1   CD shall pay to AB (i) for the preparation, production and supply of the DESIGN MANUALS the sum of £..... (pounds sterling) to be paid in instalments as follows: (due on, etc)
        and (ii) for AB's performance of its obligations assumed in Article 2.7 the sum of £..... (pounds sterling) to be paid as follows:

3.2   Payment of each instalment of the sums specified in Article 3.1 shall be made not later than 30 days following the due date. AB will provide an invoice to CD in such form as may reasonably be required.

3.3   The sums specified in Article 3.1 are net of all taxes, levies and other payments levied or payable in Country Y on such sums. Taxes, levies and payments aforesaid payable in Country Y shall be borne and paid by CD except to the extent that AB shall notify CD in writing that AB can take credit for them against its taxes in the UK on its corporate income. In that case, CD may withhold from the payment of the principal sum an amount equal to the notified available credits, and shall promptly provide to AB all required certificates and information necessary to enable AB to obtain credit therefor in the UK for the amount withheld.

3.4 The routing for payments under this Article shall be as follows: ....
or such alternative routing as AB shall reasonably substitute by notice in writing.

ARTICLE 3: COMMENTS

3.3 Tax advice essential in framing this clause.

ARTICLE 4

4.1 The contents of the DESIGN MANUALS constitute TECHNICAL INFORMATION as defined in the Licence Agreement and, accordingly, are subject to the secrecy undertakings and licences and rights defined therein.

4.2 CD will obtain from any Contractor, person, firm or State undertaking to whom any such TECHNICAL INFORMATION is supplied for the purposes of completion, construction, operation and improvement of the PLANT, a written undertaking on the terms of the Licence Agreement mutatis mutandis that they will not disclose such information to others nor use it for other purposes without the written consent of CD consistent with CD's undertakings given to AB. CD will, on request, supply to AB a copy of any such written undertakings.

ARTICLE 4: COMMENTS
Reinforcement of the Licence Agreement.

ARTICLE 5

5.1 If, at any time, it shall become apparent to AB that there is an error in the DESIGN MANUALS or that as delivered to CD or CD's nominee they were incomplete AB shall promptly remedy at its cost such deficiency and deliver duly corrected or completed versions.

Without prejudice to the guarantee terms of the Licence Agreement, the foregoing represents the entire responsibility, and liability, of AB in contract, tort, strict liability or other legal categorisation in respect of any and all such deficiencies their effects and consequences and CD will indemnify and hold harmless AB its servants and agents accordingly.

5.2 Neither party shall be liable to the other for interruption or delay in its performance of any of its obligations herein if due to causes or circumstances beyond the reasonable control of the party (including without limitation those causes normally encompassed by the term 'force majeure'). A cause or circumstance beyond the reasonable control of Messrs .... shall be deemed also to be a cause or circumstance beyond the reasonable control of AB.

Either party requiring to excuse performance or prompt performance on the aforesaid ground shall without delay advise the other party hereto with sufficient explanation of the relevant cause or circumstance and shall resume performance immediately it is able to do so. The parties will in any event consider what actions either might advisedly take to offset or lessen the effect of such cause or circumstance; any agreements made in this regard will be recorded in writing.

5.3   CD agrees that if, as a result of causes or circumstances as referred to in Article 5.2, any sum due and payable to AB is delayed in payment for more than 60 days beyond the payment date, such sum shall bear interest from the payment date at the rate of 1½% per complete month of delay compound and CD will be liable for and shall pay such interest with the delayed principal sum as soon as CD is able to do so.

ARTICLE 5: COMMENTS

5.1   Only remedy for defects is free prompt remedial design work. Second sentence legally necessary to protect Licensor.

5.2   The last sentence of paragraph 1 is important.

5.3   Assuming a substantial part of the required payments (and of the Licence fees) are advances, and recognising 2.4, the Licensor and his Contractor are not too exposed. This clause is aimed more at procedural delays than an inability to pay at all.

ARTICLE 6

6.1   This agreement shall come into effect on the date when CD shall have obtained from the relevant authorities in Country Y all necessary approvals of the terms of this Agreement, including authority to incur the obligations of payment set out herein. This date shall be promptly notified by CD to AB.

6.2   This Agreement shall be governed by the laws of Country Y. Any dispute or controversy concerning the construction of this Agreement or either party's performance of it shall, if not amicably settled by the parties, be submitted for arbitration under the rules of conciliation and arbitration of the International Chamber of Commerce by one or more arbitrators appointed in accordance with those rules. The place of arbitration shall be (as for Licence Agreement).

SCHEDULE 1

SCHEDULE 2

ACCEPTED AND AGREED	ACCEPTED AND AGREED
By:	By: ....

APPENDIX C: ASSISTANCE AGREEMENT

THIS AGREEMENT is made the ..... day of .... One thousand nine hundred and ..... BETWEEN AB Company of .... (hereinafter called 'AB') of the one part and CD Incorporated of .... (hereinafter called 'CD') of the other part.

WHEREAS CD is proposing to build a plant at .... for the manufacture of Compound X (hereinafter called 'the PLANT') and has requested that AB provide experienced people to assist and advise CD in the planning of its investment and at the commissioning and initial operation of the PLANT which AB has agreed to on the terms and conditions hereinafter appearing.

NOW IT IS HEREBY AGREED as follows:

1.   AB will second to CD a senior project manager, experienced in the planning and execution of major chemical plant investments to assist CD under secondment to plan its proposed investment, to identify and procure CD's resource requirements, to determine the basis of design for the CD PLANT and associated facilities and utilities supplies, to select and decide the proper division of responsibilities between the third party organisations required to furnish designs, detailed engineering, procurement services, and construction services, and to manage effectively all contracts placed during their execution.

2.   AB will make available to CD:
(i)   During construction of the PLANT, one or more qualified persons as determined by AB and accepted by CD (such acceptance not being unreasonably withheld), to externally inspect the plant as it is being constructed for general

compliance with the DESIGN MANUALS furnished by AB and to report to CD any evident deviations or any other omissions or errors which come to his/her attention during such inspection. CD agrees to correct or cause to have corrected such deviations, omissions or errors prior to introduction of feedstock to the PLANT. The timing of visits and the lengths of individual visits shall be discussed and agreed by the parties but no such person shall be required to be present in Country Y for more than ..... days in any calendar year.

(ii) A sufficient number of personnel possessing suitable qualifications and experienced in Compound X plant operation, as a pre-start-up team to advise on the planning of, preparations for, and the execution of commissioning and initial operation of the PLANT. Each of these personnel will be available to CD for a period of up to [six] months and the period of their assignment shall begin at least ..... weeks ahead of the planned date for the start of commissioning. The periods of assignment of these personnel may be extended by agreement between AB and CD.

(iii) Managerial personnel, if reasonably required, additional to those mentioned in (ii) above to provide CD with additional commissioning advice. The period of assignment of these additional personnel shall commence when the PLANT is close to mechanical completion but the actual date will be agreed between AB and CD in due course. The selection of the personnel shall be made by AB, and AB will make this selection bearing in mind the needs of the commissioning.

Any personnel assigned under this sub-Clause (iii) will be available to CD for a period of up to [four] months although this period may be extended in each case by agreement between AB and CD.

3.   In consideration for AB's provision of personnel as set out in Clause 1 above CD shall pay AB as follows:
(i)   A fee on the following scale:
Senior Manager or
Senior Specialist Manager: Scale A £ ..... per day
Plant Engineer/Plant Manager/
Instrument Engineer/Manager: Scale B £ ..... per day

The above daily rates shall be payable from the time that each person leaves his place of business in the UK in order to travel to Country Y until his return to his said place of business at the end of his assignment under the provisions of this Agreement save that, for each whole calendar week spent in Country Y, only 5 days shall be charged for.

(ii) All travel expenses of each person between his home in the UK and his place of residence in Country Y and all other travel expenses incurred by him whilst carrying out his duties under this Agreement (including approved periods of leave as provided for in Clause 8 below).

4.   All payments provided for in Clause 3 above will be made by CD to AB in Sterling in London within thirty days of the date of AB's invoice, which shall be rendered monthly to CD showing the number of man/days charged and any other items payable by CD hereunder. Such payments shall be net of all taxes, levies and similar payments payable on such sums in Country Y which shall be borne and paid by CD.

5.   It will be the responsibility of CD to obtain permission from the appropriate authorities in Country Y to enable CD to enter into this Agreement so as to be irrevocably bound by its terms and to enable CD to remit payment to AB in Sterling as provided for in Clause 4 above.

6.   CD will obtain all necessary work and other permits and authorisations required to enable assigned personnel and secondees to work at the Plant.

7.   AB shall have the right to replace any assigned person in the event of illness, injury, death or personal circumstances necessitating return to the UK. In this event AB shall immediately discuss the situation with CD and will ensure as far as is reasonably possible and practicable that the replacement is a person of equal qualifications and experience to the person being replaced.

8.
(i) CD will provide suitable accommodation as agreed with AB and daily subsistence for all assigned personnel. Each assigned person will have the right to return to the UK for a period of ..... days (apart from days spent travelling to and from Country Y) once in every ..... weeks PROVIDED ALWAYS that AB and CD will so organise matters to the best of their respective abilities that the job in hand is not materially disturbed by this arrangement.

(ii) In the case of any long term secondee who has his wife and family with him in Country Y, CD will pay (either to AB or to the secondee direct, as AB and CD shall agree) a daily accommodation and subsistence allowance on a scale to be agreed between AB and CD.

9. In addition to the provisions of Clause 8 above, CD will also provide for each assigned person and secondee (and his wife and family, as appropriate):
(i) Suitable and sufficient office accommodation, telephone, telex, interpreter and secretarial services.
(ii) All necessary medical assistance.
(iii) Suitable transport facilities for personal use in Country Y.
(iv) Return paid passage to the UK in the event of major injury, sickness or death.

10. Subject to the exigencies of the job, no assigned person or secondee will normally be expected to work in excess of forty hours per week, but no extra payment will be made by CD to AB in respect of any time over forty hours a week that may be worked.

11. CD will grant to each assigned person and secondee full and free access to the PLANT, to all technical information, plant data and other necessary information relating to the job and each assigned person and secondee shall have the right to be present at plant technical meetings and conferences relevant to his activities. However, no such person shall have the right to CD's cost or financial data relating to the PLANT.

12. Assigned personnel/secondees will be so chosen by AB that in AB's considered judgement they will be competent to provide adequate advice and guidance to CD but AB shall not be responsible for, and shall incur no liability whatsoever in respect of any loss, injury or damage that may result directly or indirectly from their provision of advice and guidance. AB assigned personnel will not direct the commissioning of the PLANT, or perform any executive or line-management function in CD in relation to the PLANT or otherwise.

13. This Agreement shall commence as from the date when all official approvals necessary for it to take effect have been given and shall continue thereafter until terminated by either party giving to the other not less than thirty days prior written notice of termination, such notice, however, not to be given to terminate this Agreement before .....

14. This Agreement shall be construed in all respects in accordance with Country Y law.

AS WITNESS the hand of .... duly authorised, on behalf of .... and the hand of .... duly authorised, on behalf of ....

COMMENTS

The purpose of the Assistance Agreement is to assure success and mutual satisfaction. It contains much important detail, thought to be self-explanatory. The Secondee (Clause 1) is a key person. He works for CD. Clause 12 is important.

The duration of stay of secondees and temporarily assigned experts needs to strike a balance between Project/Plant needs and tax considerations, especially avoidance of Licensor being assessed as doing business in the overseas territory. Tax levels on payments for assigned experts can be high in given instances. This agreement calls for specific net of tax payments.

The date (or event) controlling Clause 13 will have to be carefully chosen in the light of the need. It is intended as a back-stop rather than an early get out (see also the Licence Agreement).

APPENDIX D: LETTER TRAINING AGREEMENT

Dear Sirs,

(Introductory Remarks)

You have asked us to assist you in your training of a team of senior operating and maintenance personnel for that plant.

We are willing to provide training assistance, on the terms and conditions set out below. If these terms and conditions are acceptable to you, kindly indicate your acceptance by signing one of the duplicate originals of this letter where indicated at the foot and return it to us, whereupon this letter will become an agreement between our two companies which shall come into full force and effect on the Effective Date as defined in paragraph 12 below.

1. TRAINING PROGRAMME

We will provide a training programme for senior members of your operating and maintenance personnel at our UK Works during 19...... The programme will comprise an initial period of ..... working weeks during which training will be predominantly carried out at our Training Centre, and a subsequent period of ..... weeks when trainees will spend most of their training time at our Compound X plant witnessing shift working.

TRAINING METHOD

The training method employed by us will comprise instruction by and discussion with our personnel (using lectures, talks and review sessions) and exposure of your personnel to learning situations by witnessing, at first hand, actual operations and working practices being carried out on our plant.

The English language will be used throughout. You will provide interpreters if required.

TRAINING CONTENT.

In content, the training programme will cover operational and maintenance practices in our experience for all phases of normal operation of a Compound X plant, including routine practices and practices adopted by us in those upset conditions that we have found tend to be a feature of normal operations of such a plant. In addition to planned start-up and shut-down, aspects of emergency maintenance will be dealt with. In Appendix 1 are listed topics to be included in the training programme. We will consider including other topics at your request given reasonable advance notice.

To assist you in the preparation of your personnel for the Training Programme and to help identify aspects of training which might need emphasis during the Training Programme, we will make available to you at ..... a senior member of our Works Staff for a period not exceeding ..... days.

However, we wish it to be understood and agreed by you that the content of the training programme and the depth of consideration of its component topics are entirely matters for determination by us.

2.	Nothing in this letter shall be construed as placing on us any obligation to interrupt or modify the operation of any plant or to carry out any maintenance work on such plant, whether to demonstrate any operational or maintenance practice or otherwise; neither shall we be prevented from so doing. At all times the requirements of our Works management in respect of plant, personnel and other resources shall take precedence over the provision of training assistance by us pursuant hereto.

3.	We will provide for the training programme personnel who are qualified and experienced in matters covered by the training programme to the standard we expect of our own Works staff. Should our Works management decide to withdraw any person assigned to provide training for other duties, he will be replaced by another of equal qualification as promptly as practicable. We will take all reasonable steps to minimise the overall disruptive effect of such withdrawal, including, if required, extending the training programme beyond the aforesaid weeks duration. Similarly, if our plant shall have to be shut down or shall suffer abnormal upset conditions during the period of attendance by your personnel for the training programme, and as a result we are prevented from providing to them sufficient opportunity, as planned, for them to witness actual normal plant operations and maintenance, we will afford to them as soon as practicable after normal operations of the plant are resumed an equivalent opportunity without further charge.

We will not be liable to you for any extra costs or expense incurred by you by reason of any prolongation or interruption of the training programme, howsoever caused.

4. A planned date for commencement of the initial period of the training programme will be agreed before ..... but the planned date may be altered by us to a new date acceptable to you in the event of unexpected changes in the operational schedules of our plant or assignment of personnel desired by us for the provision of the training assistance temporarily to other duties.

5. CONTENT OF CD TEAM
Subject to any subsequently agreed changes, the number of CD personnel included in the team attending the initial ..... week period of the training programme should be ..... and in the following categories:
Production Manager
Shift Superintendents (the senior operating men on shift)
Process Shift Engineers
Senior Engineer responsible for Maintenance
Analytical Specialist
Materials Specialist
Instrument Specialist

CD personnel attending the subsequent ..... week period of the Training Programme will be operating and maintenance personnel selected from the above.

6. The normal working hours for day staff at our UK Works are from 8.50 am to 5.00 pm, Monday to Friday, and the initial ..... weeks of the Training Programme will be provided within these working hours. CD personnel attend-

ing shift working during the subsequent ..... week period of the Training Programme must, in the interests of safety, be fluent in spoken and written English.

7.   As consideration to us for the provision of training assistance to pursuant to this letter agreement, you will pay to us in the UK the following sums:
(i)   a fee of £..... (pounds sterling) due on the Effective Date of this letter agreement and payable within 21 days thereafter;
(ii) a fee of £..... (pounds sterling) due on the date on which the training programme specified in paragraphs 1 to 6 commences, and payable within 10 days thereafter.

These fees are exclusive of any UK VAT which may be chargeable under UK law.

No part of fees paid to us under this paragraph 7 shall be refundable to CD in any circumstances.

We will bear the 'round-trip' travel, accommodation, and subsistence expenses of our personnel who travel to Country Y pursuant to this Agreement. Similarly, you will bear such expenses of personnel travelling to the UK.

All taxes and levies outside the United Kingdom on payments due to us hereunder shall be borne and paid by you save to the extent we can obtain credit for the same under:
(a) Double Taxation Convention in force between the Governments of the United Kingdom; and
(b) any UK legislation for granting credit unilaterally.

8.   You are required to agree that you will treat as confidential all information made available to or otherwise received by you pursuant to this letter agreement (including the training programme and attendance at any plant), which we disclose to you in writing and marked confidential or which we give orally and confirm in writing within 15 days of such oral disclosure as being confidential information. You shall not disclose any information so received from us to any third party without our written consent, save that you may make any disclosure that is reasonably necessary to any consultant, or engineering firm or contractor engaged by you to perform services for you in respect of your Compound X plant, provided such consultant, firm or contractor first undertakes to treat all information so passed to him as confidential to no less an extent and

for no less a period as then applies to you in respect of that information, and to use the information passed to him for no purpose other than his performance of services for you.

The above undertakings shall not apply beyond a period of fifteen years from the Effective Date and shall not apply to information already published or which hereafter becomes published, other than by default, to information already possessed by you prior to its receipt from us, or to information acquired by you from a third party at any time without any obligation of non-disclosure and/or restriction of use imposed by said third party.

9. We will provide the aforesaid training programme in good faith but we can give no warranty that the training programme will be adequate or suitable for your purposes. The responsibility for any application by you of information received by you through the training programme or by the witnessing of plant operations and practices, and any consequences thereof, shall entirely rest with you. Accordingly, we, our servants and agents shall have no liability to you, your servants or agents in contract or in tort in respect of any services performed or any information furnished by us, our servants or agents pursuant to this letter agreement, or in respect of any subsequent application or use made by you of such information and you shall indemnify us accordingly.

10. No right under any patent or patent application is conferred by this letter agreement.

11. We will have in force Employer's Liability Insurance covering death or injury to any CD person attending our plant or Works pursuant to this letter agreement. All CD personnel entering on our Works will be required to acquaint themselves with and observe all applicable Works Rules and Safety Regulations, copies of which will be supplied in advance.

12. The Effective Date of this letter agreement shall be the date on which you receive all permissions and authorisations from the appropriate Country Y authorities necessary to enable you to enter into this agreement so as to be validly and irrevocably bound by its provisions and to remit to us all sums payable hereunder.

13. Following the completion of the Training Programme at our UK Works, all matters arising will be dealt with by letter or telex addressed to agreed

named people. However, we will have no duty hereunder to respond to requests from you for advice, information, comment or explanation concerning the subject matter dealt with in the Training Programme beyond a date ..... months following its completion.

Finally, we wish to record that we should be very pleased to welcome here two of your senior managers having responsibility for your Compound X Project and for the personnel who will be participating in the Training Programme for two or three days at a time of their choosing. Might we suggest that a good time would be towards the end of the final ..... weeks of the Training Programme.

Yours faithfully,
ACCEPTED AND AGREED
By ....
Title ....
Date ....
EFFECTIVE DATE:

APPENDIX 1: CONTENT OF THE TRAINING PROGRAMME TO BE PROVIDED

1. Aspects of the day to day operation of a Compound X Plant and the handling of the various upsets that occur, start-up of the plant from scratch and shut-down under planned conditions. Also included will be the normal and emergency maintenance of equipment associated with the plant.

2. A series of lectures will be provided by our Training Centre Staff and various Production/Engineering experts from our UK Works, and these will cover the following topics: ....

3. In addition, a substantial proportion of available time will be spent on the Compound X Plant either on normal hours or on shifts so that personnel will have direct experience of variation in operating conditions and upsets that occur.

4. There will also be a series of 'Review Sessions' where problems arising from the on-plant experiences (and other general topics) can be thoroughly discussed with our staff.

COMMENTS
The Letter Training Agreement has proved a most useful agreement on many occasions. It is a simple, workmanlike arrangement that balances needs with practicality and essential priorities. It is thought to be entirely self-explanatory.